In a teeming archive of travelogues, patriotic lyrics, and romances, *Space, Utopia and Indian Decolonization* finds in the hardness of land a dream, and in the airy imagination the hard resolve of India's postcolonial realization. In this spirited study, roving from British novels on the Mutiny, Bengali Muslim feminist fiction and the classics of Tagore, Banerjee observes that Indian subjects and British rulers alike found, even under the Raj, a utopian landscape that prefigured independence. A finely rooted yet not nationalist sally into the wonders of geographical utopia.

Timothy Brennan, *University of Minnesota*

Space, Utopia and Indian Decolonization

The book illuminates the spatial utopianism of South Asian anti-colonial texts by showing how they refuse colonial spatial imaginaries to re-imagine the British Indian colony as the postcolony in diverse and contested ways. Focusing on the literary field of South Asia between, largely, the 1860s and 1920s, it underlines the centrality of literary imagination and representation in the cultural politics of decolonization.

This book spatializes our understanding of decolonization while decoupling and complicating the easy equation between decolonization and anti-colonial nationalism. The author utilises a global comparative framework and reads across the English-vernacular divide to understand space as a site of contested representation and ideological contestation. He interrogates the spatial desire of anti-colonial and colonial texts across a range of genres, namely, historical romances, novels, travelogues, memoirs, poems, and patriotic lyrics.

The book is the first full-length literary geographical study of South Asian literary texts and will be of interest to an interdisciplinary audience in the fields of Postcolonial and World Literature, Asian Literature, Victorian Literature, Modern South Asian Historiography, Literature and Utopia, Literature and Decolonization, Literature and Nationalism, Cultural Geography, and South Asian Studies.

Sandeep Banerjee is Assistant Professor of English at McGill University. His research focuses on Anglophone, Bengali, and Hindi literature from colonial and postcolonial South Asia; Bengali and Hindi film; and literary and social theory. His articles have been published (or are forthcoming) in *Modern Asian Studies*, *Victorian Literature and Culture*, *Mediations*, and *New Global Studies* in addition to several anthologies. A recipient of grants from the Fonds de recherche du Québec–Société et culture (FRQSC) and the Gerda Henkel Stiftung in Germany, he is currently researching the ecocritical imagination of the colonial Himalaya.

Routledge/Edinburgh South Asian Studies Series

Series Editor: Crispin Bates and the Editorial Committee of the Centre for South Asian Studies

Edinburgh University, UK

The *Routledge/Edinburgh South Asian Studies Series* is published in association with the Centre for South Asian Studies, Edinburgh University – one of the leading centres for South Asian Studies in the UK with a strong interdisciplinary focus. This series presents research monographs and high-quality edited volumes as well as textbooks on topics concerning the Indian subcontinent from the modern period to contemporary times. It aims to advance understanding of the key issues in the study of South Asia, and contributions include works by experts in the social sciences and the humanities. In accordance with the academic traditions of Edinburgh, we particularly welcome submissions which emphasise the social in South Asian history, politics, sociology and anthropology, based upon thick description of empirical reality, generalised to provide original and broadly applicable conclusions.

The series welcomes new submissions from young researchers as well as established scholars working on South Asia, from any disciplinary perspective.

Agency and Knowledge in Northeast India
The Life and Landscapes of Dreams
Michael Heneise

Provincial Globalization in India
Transregional Mobilities and Development Politics
Edited by Carol Upadhya, Mario Rutten and Leah Koskimaki

Global Health Governance and Commercialisation in India
Actors, Institutions and the Dialectics of Global and Local
Edited by Anuj Kapilashrami and Rama V. Baru

Space, Utopia and Indian Decolonization
Literary Pre-Figurations of the Postcolony
Sandeep Banerjee

For a full list of titles, please see: www.routledge.com/asianstudies/series/RESAS

Space, Utopia and Indian Decolonization

Literary Pre-Figurations of the Postcolony

Sandeep Banerjee

Routledge
Taylor & Francis Group
LONDON AND NEW YORK

First published 2019
by Routledge
2 Park Square, Milton Park, Abingdon, Oxon OX14 4RN

and by Routledge
52 Vanderbilt Avenue, New York, NY 10017

Routledge is an imprint of the Taylor & Francis Group, an informa business

British Library Cataloguing-in-Publication Data
A catalogue record for this book is available from the British Library

Library of Congress Cataloging-in-Publication Data
Names: Banerjee, Sandeep, author.
Title: Space, utopia and Indian decolonization : literary pre-figurations of the postcolony / Sandeep Banerjee.
Description: Abingdon, Oxon ; New York : Routledge, 2019. | Series: Routledge/Edinburgh South Asian studies series | Includes bibliographical references.
Identifiers: LCCN 2018056635 | ISBN 9781138393684 (hardback) | ISBN 9780429401626 (ebook) | ISBN 9780429686399 (epub)
Subjects: LCSH: South Asian literature—History and criticism—Theory, etc. | Space and time in literature. | Colonies in literature. | Nationalism in literature.
Classification: LCC PK5407 .B36 2019 | DDC 895—dc23
LC record available at https://lccn.loc.gov/2018056635

ISBN: 978-1-138-39368-4 (hbk)
ISBN: 978-0-429-40162-6 (ebk)

Typeset in Times New Roman
by Apex CoVantage, LLC

MIX
Paper from responsible sources
FSC® C013056

Printed and bound in Great Britain by
TJ International Ltd, Padstow, Cornwall

Contents

Acknowledgements viii
Note on the use of Bengali, Hindi, and Sanskrit x
List of figures xii

Introduction: spatial desire in the age of empire 1

1 Of good and evil: the anxiety of utopianism 18

2 Tales of a city: writing colonial Calcutta 51

3 That magnificent song: between the performative and the pedagogic 82

4 A sense of place: narrating knowable communities 116

Epilogue: that im/possible spatial desire called decolonization 148

Appendix: Jana Gana Mana *by Rabindranath Tagore* 154
Bibliography 157
Index 172

Acknowledgements

This book has collected numerous debts since it began its journey as a set of scribbled notes in Syracuse, New York. I am grateful to Crystal Bartolovich, Don Mitchell, Subho Basu, Don Morton, and Kevin Morrison for helping me shape my initial thoughts on literature, space, utopia, and de/colonization in their distinct and distinctive ways.

Multiple other extensive and extended conversations over the last few years, conducted in person and virtually, have helped me immensely to recast my earlier thoughts and transform them into this book. I thank my wonderful community of interlocutors in Montréal and beyond: Subho Basu remains a wonderful teacher, colleague, and collaborator from whom I continue to learn. For helping me think through my ideas as well as their solidarity and comradely critiques, Auritro Majumder, Prashant Keshavmurthy, Koel Banerjee, Meghant Sudan, and Gavin Walker. For their support and encouragement, Keya Ganguly, Henry Schwarz, and Timothy Brennan. I remain grateful to Amlan Das Gupta for many things, only one of which is his academic mentorship.

For helping me locate my research materials, I thank the staff of the Humanities and Social Science Library of McGill University; the E.S. Bird Library of Syracuse University; the British Library and the India Office Records; the National Library of India. I gratefully acknowledge the help with images and documents I received from Subhashish Roy, Arijit Sen, Anirban Acharya, and Shiladitya Banerjee. I also thank Crispin Bates, the Series Editor of the Routledge/Edinburgh South Asian Studies Series as well as the editors at Routledge, Dorothea Schaefter and Alexandra de Brauw.

I am grateful to Sofia Cutler and Felix Fuchs for their research assistance. I also thank Suvij Sudershan for his help with research and copy-editing of this manuscript. I am grateful to Sunil Sharma for help with locating Ghalib's poetry on Calcutta, and to Prashant Keshavmurthy for translating some of it into English from the original Farsi.

I have had the unenviable experience of having to deal with several health-related setbacks as I have worked on this book. For getting me back on my feet on a number of occasions without making me penurious in the process, my heartfelt thanks to the socialized healthcare of Québec, especially the doctors and the excellent nursing staff of l'hôpital Hôtel-Dieu and l'hôpital Saint-Luc of Montréal. I truly am very grateful.

I am fortunate to have wonderfully supportive colleagues in the English department at McGill. I thank Katherine Zien and Monica Popescu for their friendship as well as for their help with innumerable things, large and small, over the years. I also thank Trevor Ponech, Allan Hepburn, Miranda Hickman, Erin Hurley, Maggie Kilgour, Michael Nicholson, Derek Nystrom, Ara Osterweil, Fiona Ritchie, and Ned Schantz.

My friends sustained me with a steady stream of banter, laughter, *adda*, and *khilli*. For their companionship during the best of times and helping me to survive the worst of times, I thank Cyril Ghosh, Arijit Sen, Shiladitya Banerjee, Auritro Majumder, Pragna Sen, Sanchari Dutta, Devalina Mookerjee, Anirban Acharya, Subhashish Roy, Sunandini Banerjee, Sanjit Basu (Babu), Priyodarshi Banerjee, Debdan Banerjee, Anirban Sen, Saptarshi Sanyal, Sreya Chatterjee, Aparajita Bhattacharya, Dipyaman Sanyal, as well as my friend-in-law, Koel Banerjee.

I also thank Apurva Ashok, Anupam Basu, Gautam Basu Thakur, Paromeeta Mathur Banerjee, Arijit Barman, Sanjay Basak, Janie Bériault, Rishabh Bhandari, Laura Boothman, Arpita Chatterjee, Rimi B. Chatterjee, Soma Chatterjee, Aditi Das Gupta, Shamya Dasgupta, Niharika Dinkar, Prachi Deshpande, Prasanta Dhar, Ajitha G. S., Marie-Eve Drouin-Gagné, Jocasta Gardner, Gerard Gaskin, Raf Gelders, Ambarnath Ghosh, Tanushree Ghosh, Abhijit Gupta, Niharika Gupta, Gal Gvili, Malavika Kasturi, Rekha Krishnan, Ananda Lal, Arpan Majumder, Zain Mian, Priyanjali Mitra, Radhika Mongia, Reshmi Mukherjee, Sajni Mukherji, Shane O'Brien, Andrea Pinkney, Nimanthi Rajasingham, Shakuntala Ray, Rachel Sandwell, Gopa Sen, Sarah Nafisa Shahid, Gohar Siddiqui, Malgorzata Skorobohata, Pushkar Sohoni, Aakash Solanki, Erin Soros, Aparna Sundar, Akshaya Tankha, Kristin Voigt, Susan Wadley, Bridget Walsh, Luke Whitmore, and Xing Zhang.

I thank my parents, Shibani Banerjee and Santosh Kumar Banerjee. My mother has been a constant source of love, support, and encouragement and a model of kindness and grace. I am fortunate to have Sreerupa Roy and Susmita Roy for my cousins, and always just a phone call away. They provide me with familial joys as do Partha Roy, Utsa Roy (Coco), and Pranati Pal. I also thank my parents-in-law Paromita and Biplab Majumder for being tremendous sources of support.

My partner Atreyee Majumder put up with me as I finished the manuscript with patience and good humour even as she cautioned me against too much attention to the syntax of things. She fills my life with love and care but also theory and poetry, making it joyous in more ways than I can count or care to recount.

Asim Kumar Datta, Diptosh Majumdar, A.D. Nuttall, Pratap Mitra, and my grandmother, Sulekha Basak, would have been thrilled to know that I have published my first monograph. They are, sadly, not here to share this moment with me.

Dr. Datta introduced me to the exciting world of humanistic inquiry. Diptosh helped me to make sense of Indian politics before supporting me through my seemingly hare-brained attempt to quit television journalism mid-career and return to academics. I remember them, and our conversations, with great fondness. I recall their generosity, affection, and kindness with gratitude. I commend this book to their memory.

Note on the use of Bengali, Hindi, and Sanskrit

Space, Utopia and Indian Decolonization cites Bengali texts extensively and also references Hindi and Sanskrit phrases and terms.

All quotations from Bengali texts are in English translation. I have used English translations of Bengali texts where available or translated passages from the Bengali myself. These are acknowledged in the Notes. For passages translated by me, I have provided the original Bengali where I have deemed it necessary. My translation of Tagore's *Jana Gana Mana* is provided as an appendix.

The English titles of Bengali texts are followed by the original Bengali names in square brackets. For instance, *Anandamath* [আনন্দমঠ]. At times, English phrases or words translated from the Bengali, Hindi, or Sanskrit are followed by the phrase or word in the original language within square brackets. For instance, *impenetrable* [দুর্ভেদ্য; durbheddyo]. At other times, a word or phrase from these languages is transliterated and the original word and its meaning follow. For instance, "The Bengal famine, called *chhiyattorer monwontor* [ছিয়াত্তরের মন্বন্তর] in Bengali, meaning 'the famine of '76' after the year 1176 of the Bengali calendar . . ." This is aimed to provide readers who are familiar with these South Asian languages a sense of the original.

Bengali texts consulted in the original language are cited in the Bibliography in English transliteration with the Bengali following in square brackets. For instance, Chattopadhyay, Geeta. *Bangla Swadeshi Gaan* [বাংলা স্বদেশী গান]. Delhi: Delhi University, 1983.

I have used the English spellings for Bengali names and texts from available translations even though they do not conform to the actual Bengali pronunciation. For instance, *Anandamath* instead of *Anondomoth*; *Jana Gana Mana* instead of *Jono Gono Mono*; *Nanda* instead of *Nondo*; and *Padmarag* instead of *Poddoraag*. For all other Bengali words, I have transliterated them into English keeping the original pronunciation in mind. For instance, *Bongodorshon* and not *Bangadarshan*; *bharotborsho* not *bharatvarsha*; and *jonogono-oikkyobidhayok* not *janagana-aikkyavidhayak*.

The common Bengali surname Chattopadhyay is Anglicized into Chatterjee or Chatterji. This became an issue for Bankim as both variants of the Anglicized surname have been used by publishers for his English texts. I have followed Julius Lipner's convention and used *Chatterji* when referring to Bankim throughout this book. In the Bibliography, all of Bankim's entries are filed under Chatterji. In cases where a different spelling of "Chatterji" has been used in the original published text, that spelling appears in square brackets after "Chatterji." For instance, Chatterji [Chatterjee], Bankim Chandra. *Durgesa Nandini or, The Chieftain's Daughter: A Bengali Romance*. Translated by Charu Chandra Mookerjee. Calcutta: H. M. Mookerjee & Co., 1880.

Figures

1.1 Popular depiction of the Hindu goddess Kali 40
1.2 Durga being worshipped *en famille* in South Calcutta 42
3.1 *The East Offering its Riches to Britannia*, 1778, Spiridione Roma 88
3.2 *The British Lion's Vengeance on the Bengal Tiger*, 1857 89

Introduction

Spatial desire in the age of empire

Decolonization – typically understood as the dismantling of the European empires in Ireland as well as throughout Asia, Africa, and the Caribbean – was arguably the most significant and world-historical expression of a struggle over space in recent memory.[1] It was only less than a century ago that we lived in an imperial global order aptly described by Frantz Fanon as "a world divided into compartments."[2] That was a world where *metropolitan* states, through a complex of economic, political, and cultural controls, ruled over the *peripheral* world that consisted of the greater part of the earth's surface and of the majority of the earth's residents. The processes of political decolonization transformed that unequal world into our contemporary world-system of nominally equal territorialized nation-states. This intense struggle against colonialism – its political, economic, and cultural structures and strictures – was fought, as Edward Said reminds us, not only with soldiers and cannons but *also* ideas, images, and imaginings.[3] The struggle over geographical ideas, images, and imaginings – *imagined geographies*, in sum – was staged in, and through, literary texts that made up the global literary field of the late nineteenth and twentieth centuries, that is, the one and unequal colonial literary world-system.[4] These texts partook in, and were marked by, the struggles against and for colonialism as their imaginaries sought to re-present the spaces they depicted through stances ranging from the colonial to the counter-colonial.

The literary field of the colonial world was where colonial and counter-colonial spatial imaginaries jostled with each other for hegemonic pre-eminence. It was a key site for spatial imaginings and re-imaginings and a fundamental locus for fashioning the ideologies that undergirded the larger politics of decolonization. The spatial imaginaries of these colonial and anti-colonial texts gesture towards a *spatial desire* that informs their ideological task of instantiating imagined geographies.[5] The spatial desire of colonial texts makes them posit, explicitly or in indirect ways, imagined geographies that reaffirm the hierarchical vision and division of the world. Anti-colonial texts, on the other hand, appear marked by a converse spatial desire. They seek to refuse colonial spatial conceptions and re-imagine them in a counter-colonial register. In effect, they perform a determinate negation of the spaces of the colonial present, illuminating them as limited, and as spaces of limitation. They re-imagine colonial spatial registers in a utopian vein to insinuate, in their stead, possible counter-colonial spaces: spaces of counter-colonial possibility that I call the *postcolony*.[6] The postcolony, in my use of the term, signals a pre-figuration of the decolonized space of a colony: inchoate, yet to be fully crystallized, and grasped only as a structure of feeling.[7] Let me add a caveat here: I do not think of the postcolony as a singular entity but, as I show throughout this book, a category that is decidedly plural, inherently multivalent, and intensely contested. In my formulation, spatial desires and utopias have the relationship of content and form: spatial utopias are the sedimentations of spatial desires.

Situated at the nexus of English and comparative literary studies, as well as studies on space, utopia, and decolonization, this book examines colonial and anti-colonial texts from a specific period and at a specific locale in the global literary field. *Space, Utopia and Indian Decolonization* interrogates and annotates the spatial desires of texts that make up the British Indian literary field – the literary field of colonial South Asia under British rule – between the decades of the 1860s and the 1920s to excavate their imagined geographies and illuminate how these texts posited and contested geographical imaginations. I focus on this period as it was a time *before* the "moment of arrival" of hegemonic forms of anti-colonial struggles and was instead marked by "contradictions, divergences, and differences," when the various constituents of the anti-colonial struggle in British India were working out what decolonizing British India meant, and entailed.[8] My stress on thinking about the relationship between narrative and social space draws from the pioneering work of Raymond Williams and its subsequent elaboration by Edward Said.[9] Commenting on the spatial dimension of humanistic inquiry, Said notes that "structures of location and geographical reference appear in the cultural languages of literature, history, or ethnography, sometimes allusively and sometimes carefully plotted."[10] Despite this, he goes on to comment, most "cultural historians, and certainly all literary scholars, have failed to remark the *geographical* notation, the theoretical mapping and charting of territory that underlines Western fiction, historical writing, and philosophical discourse of the time."[11] Said's critical comment anticipates what is now a small but developing body of scholarship that emphasizes *space* as a key vector of literary and cultural analysis, especially in colonial and postcolonial literary studies.[12] *Space, Utopia and Indian Decolonization* joins this critical corpus to think through colonization and decolonization as global projects of space-making and place-making, and their literary registrations.

Space-making, place-making

Human beings, Karl Marx notes, by "acting on the external world and changing it . . . at the same time [also] change . . . [their] own nature."[13] This act of interaction between human labour and external nature begins the process of the social production of external nature *and* space. Under capitalist relations of production, labour is unmoored from the process of reproducing social life and, in becoming independent of that process, transforms into an abstraction or *abstract labour*. A spatial correlate of this process is the reconstitution of place into space, that is, the abstraction from local and rooted notions of place to conceiving of "spatial extension beyond immediate experience."[14] This results in the conceptual fusion of place and society being broken, leading to what Henri Lefebvre has called the social production of space.[15] The production of space under capitalism, he notes, transforms absolute space into *abstract space*.[16] For Lefebvre, abstract space is

> a product of violence and war, it is political; instituted by a state, it is institutional . . . [and] appears homogenous; and indeed it serves those forces which make a *tabula rasa* of whatever stands in their way, of whatever threatens them – in short, of differences.[17]

Lefebvre notes the tendency of abstract space to proliferate across the globe, and this observation points to not just the space-making but also the *world-making* dimension of colonial capitalist modernity that produces the world-market and the world-economy, and indeed the contradictory unity that is our one and unequal modern world-system.[18] These world-making processes are typically referred to by the conceptual shorthand *colonization* but should, more accurately, be called *colonial globalization*, which brought about what Martin Heidegger in a fit of poetic genius labelled the "age of the world-picture."[19]

Elaborating on the production of space in a global scale, Neil Smith notes that as more and more concrete, and isolated, labour processes are made abstract and brought together in relations of exchange as wage-labor, more and more absolute spaces are integrated into the world-economy as relative and abstract space. In other words, *place* is transformed into *space*. The *universalization* of capitalism the world over also leads to the universalization of the division of labour and the concomitant inscription of *hierarchical difference* at the heart of the capitalist mode of production. This also, and necessarily, has a spatial correlate: the division of the world "into legally distinct parcels, divided by great white fences, real or imaginary."[20] The process of dividing the world into "legally distinct parcels" that Smith notes also points to the production, and indeed the emergence, of a novel spatial form that is fundamentally tied to colonial capitalist modernity: the *border*, a spatial category that structured – and continues to structure – human subjectivities in decisive ways. The colonial provenance of the border-concept is outlined by Thongchai Winichakul in his study on Siam (Thailand). Interactions between the British and the Siamese, he notes, were rife with discrepant understandings of what a border signified since for the Siamese, "the limits of a kingdom could be defined only by [border] townships' allegiance to the centre of a kingdom [and the] political sphere . . . only by power relationships, not by territorial integrity."[21] While the Siamese did not lack the terminology and concepts for boundaries, Winichakul notes that those categories tended to signify "areas, districts, or frontiers, not boundary lines. They mean[t] a limit – an extremity without a clear-cut edge and without the sense of division between two powers."[22] The development of the border-concept is coeval with those of other new spaces: new kinds and conceptions of hierarchical spaces such as the metropole, periphery, and the semi-periphery, linked together in an imbricated relationship of colonial capitalism.[23] This production of the world as a concatenation of *spaces*, namely, bounded territorial markets for the buying and selling of labour power and other commodities, also presents it as a disenchanted space of colonial capitalist modernity. To render it into a *lived space*, these spaces must be re-animated with an affective charge and re-signified as *dwellings*. The literary field, across the colonial and counter-colonial registers, is tasked with a *place-making* function: to re-enchant the disenchanted spaces of colonial capitalist modernity and provide them with a sense of *place*. In the context of South Asia, the contending forms of spatial desire seek to re-enchant a specific bounded expanse of abstract space and present it as India the colony or the postcolony. Alongside its place-making function, these representations also fulfil a pedagogic function of teaching readers about India's colonial or counter-colonial identity.

In *Space, Utopia and Indian Decolonization*, I draw out the place-making and pedagogic functions of colonial and anti-colonial texts in addition to underlining the literary registration of the spatial politics of colonialism and anti-colonialism. My aim here is twofold: I seek to *spatialize* our understanding of decolonization while at once seeking to decouple and complicate the easy equation between *decolonization* and *anti-colonial nationalism*.[24] Studies on anti-colonial nationalism, especially in South Asia, tend to collapse decolonization into nationalism with even the most insightful of them implicitly presenting the (Indian) nation as the necessary *telos* of the decolonizing process.[25] These studies, as Sugata Bose has noted, have the "unintended effect of playing into the hands of the postcolonial state's ideological project which it promises to question and critique" in addition to being tinged with an avoidable presentism.[26] Instead, by thinking with, and through, the category of the postcolony, I suggest that decolonization and its spatial registers as open, plural and contested. And by showing contending visions of the postcolony that ranges from the parochial to the universal, I most emphatically refuse the simplistic characterization that "anticolonial nationalism is seldom narrow, sectarian or chauvinistic; it seeks instead to open the community of the globe."[27] More significantly, by showing anti-colonial texts partaking in the ideological and spatial politics of decolonization, I not only delineate the utopianism of their spatial desires but also situate *literary imagination* and *representation* as fundamental to the project of decolonization. These categories are notoriously absent in studies on Indian decolonization and anti-colonial nationalism. It is thus not at all surprising to find Dipesh Chakrabarty comment that "Benedict Anderson's book *Imagined Communities* has made us all aware of how crucial the category 'imagination' is to the analysis of nationalism [but] . . . compared to the idea of community, imagination remains a curiously undiscussed category in social science writings on nationalism."[28] And while Chakrabarty speaks of social science writings on nationalism, this is particularly true of postcolonial theory and literary studies, which Neil Lazarus has charged, on several occasions, with the disavowal of decolonization.[29]

Space, Utopia and Indian Decolonization, furthermore, draws attention to the imbricated relationship of *decolonization* and *utopianism*. It posits decolonization as a concrete effect of a project of collective dreaming and the concomitant "education of desire."[30] And if utopianism is typically understood as the expression of a "desire for a better way of being and living," it spatializes this insight by thinking of the category with, and through, decolonization.[31] I suggest that the utopian horizon of decolonization was not simply, or only, the expression of a desire for a better *way* of being and living but also, and just as crucially, the desire for a better *locus* of being and living. In the colonial context therefore, the "education of desire" that is utopianism was also, and necessarily, the education of a spatial desire that we call decolonization. And anti-colonial texts partaking in this process engaged in the *utopian task* of re-imagining the colony as the post-colony even as they performed the pedagogical task of providing an *aesthetic education* about it. By thinking about the *literary* and the *utopian* together, this book illuminates the anticipatory character and function of literary representations and the literary field. This is inspired by the influential theorization of utopia by Ernst Bloch, who understands it as a kind of "anticipatory consciousness" structured by the principle of "Not Yet."[32] For Bloch, the principle of "Not Yet" has two aspects, the "Not-Yet-Conscious" and the "Not-Yet-Become," which map out its ideological and material, or subjective and objective correlates. Transposing this schema to the colonial context, we could say that the principle of "Not Yet," that is, the utopian horizon of spatial desire, is the colony or the postcolony. The literary field and its constituent literary texts, then, make up its ideological or subjective correlate – the "Not-Yet-Conscious" – whereas the processes of colonization and decolonization become its material and objective condition of possibility, the "Not-Yet-Become." To suggest the literary field as a site of the "Not-Yet-Conscious," or ideology as such, is to point to the intimations of utopia in literary texts. While the body of scholarship on literary utopias and utopianisms has conceptualized it across axes of form, content, function, and method, these have typically focused on the explicit utopian yearnings and articulations of texts.[33] Thinking about spatial desire as the "Not-Yet-Conscious" of literary texts is then to be mindful of, and to draw attention to, their *latent utopianism*. And this necessarily raises the vital hermeneutical question: *how to read texts for their latent spatial desire?*

Reading for spatial desire, globalectically

"Ideas are to objects," Walter Benjamin writes, "as constellations are to stars."[34] Benjamin's comment outlines a reading strategy, subsequently developed by Theodore Adorno, that suggests the illumination of an idea through the conscious bringing together, or the *constellation*, of objects.[35] Building on this hermeneutical method, I suggest that the latent spatial desires of texts are made visible when they are constellated with other texts and read relationally. In other words, the uneven terrain of spatial desires of anti-colonial texts are rendered more clearly legible when these texts are read in relation to their colonial counterparts. I therefore read contemporaneous colonial and anti-colonial texts together, and in a comparative vein, to understand their representations of space. My method is akin to – but significantly different from – what Said calls *contrapuntal reading*. In outlining his reading method, Said calls for an "interpretative change of perspective . . . to challenge the sovereign and unchallenged authority of the allegedly detached Western observer."[36] Urging literary and cultural critics to take western cultural forms "out of the autonomous enclosures in which they have been protected" he insists on locating them instead "in the dynamic global environment created by imperialism."[37] To read the cultural archive "contrapuntally" for Said, then, is to demonstrate "a simultaneous awareness both of the metropolitan history that is narrated and of those other histories against which (and together with which) the dominating discourse acts."[38] While contrapuntal reading does indeed point to the worldliness of texts and situates them in the dynamic historical context of colonial globalization, Said's critical agenda nevertheless remains focused on the *western* cultural archive and rooted in a logic of canonicity.

In *Space, Utopia and Indian Decolonization*, I move beyond the categories of the "western" and hope to pressure the label of the "canonical." To begin with, I take up the recent exhortation by Jacqueline Dutton and Lyman Tower Sargent to examine notions of utopia and utopianism in "non-western" cultural traditions.[39] This allows me to move the critical conversation in the field of utopian studies away from the almost exclusive focus on the western canon.[40] More crucially, I constellate western or metropolitan texts with their non-western or peripheral counterparts. Specifically, I constellate British literary texts from, largely, the Victorian era and Bengali literary texts that co-inhabited in the literary field of colonial South Asia fashioned by global capitalism under the sign of colonialism. In so doing, my method seeks to *globalize* Bengali literary texts situating them as moments of world literature even as it *provincializes* English (Victorian) literary texts.[41] This allows me to make legible the non-west – South Asia, or Bengal, to be precise – as legitimate agents of literary and cultural production, while also challenging the phenomenon Chakrabarty has aptly termed "asymmetric ignorance" where non-western scholars are required to know metropolitan history, literature, and theory without any converse expectation of metropolitan scholars.[42]

In addition, and just as significantly, I read across the English-vernacular divide to understand space as a site of contested representation, and ideological contestation. My atypical critical agenda draws from the conception of comparative reading developed and outlined by Ngũgĩ wa Thiong'o that he calls *globalectics*. I read these literary texts in a global comparative framework – *globalectically* – to highlight the uneven organization of the world-literary space even as it seeks to transcend that unevenness by stressing the "equality of potentiality of parts."[43] Such a reading strategy, moreover, pushes against the national organization of the literary canon as well as the methodological nationalism that dominates literary and cultural studies scholarship around the globe. And this is particularly apposite for the South Asian literary field that has been a literary multiverse through the pre-colonial and colonial periods, and continues to be so in the postcolonial era. The literary traditions of several South Asian languages – Bengali, Urdu, Tamil, Punjabi, among others – not only refuse to be contained within contemporary national boundaries but have historically interacted with each other in organic ways.[44] The Anglophone literary tradition of South Asia, which is marked by specific class and caste – and in the colonial era, race – markers, complicates this picture further. For instance, the Bengali novel which emerges during the colonial era through interactions with, and as a response to, the British novel in turn serves as the model and interlocutor for novels in Hindi and other South Asian languages.[45] Any serious consideration of the literary field of colonial (or postcolonial) South Asia must therefore necessarily adopt a comparative perspective that goes beyond the limited, and limiting, focus on the region's Anglophone literary universe and engage with literary works produced in the South Asian vernaculars.[46] And this is something that English-language scholarship on South Asian literary texts, especially in the field of postcolonial literary studies, has failed to do.[47] My reading strategy, which brings together British and British Indian texts as well as texts written in English and Bengali, then, makes evident the *worldliness* of these texts, showing them as members of a global literary field whose spatiotemporal relationship is structured by colonial globalization. Let me illustrate this method with some concrete examples.

Rabindranath Tagore's poem from the 1901 anthology *Noibedyo* [নৈবেদ্য], offers a vision of a future for India and the world. He writes:

> Where the mind is fearless, the head held high
> And knowledge, free. Where, day and night,
> This earth is not broken by walls,
> Not turned into tiny, fragmented parts.
> Where speech is heart-felt. Human actions,
> Seeking fulfilment across lands,
> In all directions take flight.
> Where the sands of petty habit
> Do not choke reason's streams;
> Humanity not shattered into myriad bits.
> Where you, night and day,
> Shape thought, work and play.
> Be unsparing, my Lord – and strike!
> To awaken India into that paradise.[48]

In the poem, Tagore presents India, and the world, in a state of becoming that appears to be marked by Bloch's utopian principle of the "Not Yet." He characterizes the future as a site – "that paradise" – with specific attributes. This is an image of India where the mind is without fear, the head is held high, and knowledge, free. The world is also envisioned as one, and not broken into parts by walls; a place where the heart shapes speech, directed human action seeks fulfilment across the globe, and a place where reason has not lost its way in the petty sands of habit. This representation of the future is coded with a spatial desire that is explicitly utopian. It *seeks to stretch India* to make it more than itself, and a metonym of the world.

In contradistinction to Tagore's view of space is Vinayak Damodar Savarkar's. Writing about India in a tract, the doyen of Hindu Right thought in South Asia notes: "this Bharatbhumi [India], this Sindusthan, this land of ours that stretches from Sindhu to Sindhu is our Punyabhumi [holy land], for it was in this land that the Founders of our faith and the Seers to who 'Veda' the Knowledge was revealed."[49] In this passage, Savarkar collapses the identities of India and Hinduism into each other, privileging the identity of the former as a space of Hindu revelation. This allows him to make claims about who does, or does not, belong to India elsewhere in the same tract:

> So to every Hindu, from the Santal to the Sadhu this Bharata bhumi [India] this Sindustan is at once a Pitribhu and a Punyabhu – fatherland and a holy land.
>
> That is why in the case of some of our Mohammedan or Christian countrymen . . . [who] have inherited along with Hindus, a common Fatherland and a greater part of the wealth of a common culture – language, law, customs, folklore and history – are not and cannot be recognized as Hindus. For though Hindustan to them is Fatherland as to any other Hindu yet it is not to them a Holyland [sic] too. Their holyland [sic] is far off in Arabia or Palestine. Their mythology and Godmen, ideas and heroes are not the children of this soil. Consequently their names and their outlook smack of a foreign origin. Their love is divided.[50]

Savarkar's depiction of India indicates the concretization of another kind of spatial desire. His India is a land of myth: it is a holy land and a fatherland, and its true inhabitants are those who acknowledge it as such, that is, the Hindus. Savarkar excludes some of its inhabitants from the space of India. In so doing, he *seeks to reduce India,* making it less than itself. Tagore and Savarkar delineate two distinct conceptions – two diametrically opposed conceptions, in fact – of Indian space. But my larger point here is not so much about the *content* of their imagined geographies but rather one of *method.* It is only by reading these two spatial imaginaries in relation to each other that we see them as divergent and contestatory of each other.

Significantly enough, neither of the two spatial imaginations of India appear to be counter-colonial in any obvious way. But their anti-colonialism, that is to say, their counter-colonial spatial desire becomes evident when we juxtapose these with a normative colonial conception of India. For such a view, we could turn to the colonial bureaucrat John Strachey:

> What is India? What does this name India really signify? The answer that has more than once been given sounds paradoxical, but it is true. There is no such country, and this is the first and most essential fact about India that can be learned.
>
> India is a name which we give to a great region including a multitude of different countries. There is no general Indian term that corresponds to it. The name Hindustan is never applied in India, as we apply it, to the whole of the Indian continent; it signifies the country north of the Narbada [Narmada] River, and especially the northern portion of the basins of the Ganges [Ganga] and Jumna [Yamuna].[51]

Strachey's claim that there was no India speaks to a colonial conception of the space. Indeed, it spatializes the colonial desire that sees India as *terra nullius*, or an empty space that is given content and form by colonialism. This allows Strachey, elsewhere in the text, to reassert his claim there is no such category as India, or the "people of India":

> This is the first and most essential thing to learn about India – that there is not, and never was an India, or even any country of India, possessing, according to European ideas, any sort of unity, physical, political, social, or religious; no Indian nation, no 'people of India,' of which we hear so much.[52]

Constellating Strachey's conception of India with those of Tagore and Savarkar lights up the counter-colonial spatial desire of the latter two. It shows how both the imagined geographies posited by the texts of Tagore and Savarkar, though in fundamentally different ways, seek to contest colonial conceptions of space, and reimagine the colony as the postcolony. These representations, taken together, point to a struggle within the colonial literary field about conceptions of Indian space and belonging. Each of these representations seek to present India as an imagined geography and is in quest of hegemony. By providing the signifier "India" with specific kinds of content, these representations seek to make it legible as a colony or a postcolony. At the same time, these representations are invested with the task of cultural pedagogy: of teaching their readers *what India is* and, just as significantly, what *India may be*.

In the chapters that follow, *Space, Utopia and Indian Decolonization* interrogates the spatial desire of anti-colonial and colonial texts across a range of genres, namely, historical romances, novels, travelogues, memoirs, poems, and patriotic lyrics. This literary archive is at times supplemented by an array of non-literary sources: reports of various departments of the British Indian government, newspaper articles, colonial travel guides, popular geography books, and colonial ethnography. Through detailed and nuanced textual analyses, it underlines that while all counter-colonial spatial imaginaries are utopian in their negation of colonialism, they are, nevertheless, radically different from – and at times mutually incompatible with – each other. By showing that counter-colonial imaginaries range from the parochial to the universal, it pluralizes the categories of the "anti-colonial" and the "postcolony" underlining them as fundamentally conflicted and multivalent.

In the first chapter, I examine two colonial romances from British India: Philip Meadows Taylor's "mutiny novel" *Seeta* (first published in 1872) and Bankim Chandra Chatterji's Bengali novel *Anandamath* [আনন্দমঠ] (first published in 1882). Drawing on theorizations of the romance that suggest the form lays out the antinomy of good and evil, I examine these novels' articulations of utopian spaces and subjects. *Seeta* presents the eponymous heroine Seeta and the colonial official Cyril Brandon as epitomes of good while Azarel Pande embodies the principle of evil. It also extends this antinomy of good and evil to its spatial imaginaries investing them with moral qualities thus positing visions of utopia and dystopia. I also show that the narrative suffers from anxieties about the inter-racial marriage of Seeta and Brandon as well as the existence of the mutinous subject Azrael Pande. These anxieties are resolved by killing off the characters, thereby effecting a kind of social catharsis that transforms the narrative world in a utopian vein. Chatterji's *Anandamath* codes the narrative's two co-existing spatialities – the ahistorical space of the forest and the famine-afflicted village – with the good-evil antinomy where the forest signals, in an anachronistic vein, the future space of the postcolony in the present. The novel also deploys the good-evil antinomy to characterize social groups, namely, Hindus, Muslims, and the British. The utopian horizon of Chatterji's text is the decolonization of Indian space but the attempt to narrativize the overthrowing of British rule creates an impossible vision of decolonization, and a crisis for the narrative. This is resolved, I show, through the use of the *deus ex machina* at the novel's denouement that not only defers the end of British rule to an indeterminate future but also acquiesces to the logic of colonial progressive history. Chapter 2 situates Calcutta, the capital city of British India (until 1911), in the confluence of diverse articulations of spatial desire in a range of literary and non-literary texts from the nineteenth and early-twentieth centuries. My literary archive includes poems, travelogues, histories, and memoirs by writers such as James Atkinson, Mirza Ghalib, Bholanath Chunder, Rudyard Kipling, Binay Krishna Deb, and Rabindranath Tagore, which is supplemented by government reports, tourist guidebooks, popular geography texts, and newspaper reports from *The Statesman*, *Englishman*, and *The Pioneer*. Understanding the city, following Lefebvre, as an *oeuvre*, I explore how these depictions of Calcutta and its residents inscribe the urban landscape with widely differing meanings at different moments in its colonial history. And I illustrate the contested ways of perceiving, conceiving, and being in the colonial city by reading Calcutta as a scaled-down version of the post/colony.

In the next chapter I make a case for understanding the patriotic lyric as a global as well as a performative and pedagogic genre. It not only teaches its auditors and readers about space and belonging but also, owing to its performative nature, enacts the idea of spatial belonging. I outline features of the patriotic lyric by focusing on metropolitan patriotic lyrics such as James Thomson's "Rule! Britannia" (composed in 1740) before examining two patriotic lyrics from colonial India: Bankim Chandra Chatterji's *Vandemataram* [বন্দেমাতরম] (first published as a poem in the journal *Bongodorshon*, in 1875; subsequently published as a part of *Anandamath*, in 1882) and Rabindranath Tagore's *Jana Gana Mana* [জনগণমন] (published in 1911). The lyrics, whose truncated versions are the National Song and National Anthem of the postcolonial Indian state, both structure their spatial imagination through landscapes to imagine India as a postcolony. Despite the formal similarities, however, the lyrics articulate two distinct, and divergent, conceptions of Indian space. Chatterji's lyric presents Indian landscape in a pastoral vein that re-imagines the contemporaneous lived space of famine and death in a utopian vein but nevertheless evacuates all traces of labour from it. This re-imagination re-enchants the landscape by deifying it as a mother goddess of Hindu religious traditions and posits Bengal (and India) as a Hindu majoritarian space. While Tagore also deploys the landscape form in *Jana Gana Mana*, the vision, however, is plural. He presents Indian space as distinct from divinity, and a palimpsest of its physical, political, and ethnic units, and as an inclusive civilizational space for all communities. I also delineate *Jana Gana Mana* as a counter-pastoral that makes visible the role, and labour, of India's ethnic and religious communities in shaping and producing that space besides attempting to find a plural and common idiom for narrating Indian space and its communities. In the final chapter, I take up the following questions: how do novels from colonial India articulate notions of community? And how do they gesture towards the idea of belonging to Indian space? I examine these questions through a constellatory reading of Rudyard Kipling's *Kim* (first published in 1901), Rabindranath Tagore's *Gora* [গোরা] (first published in 1909), and Rokeya Sakhawat Hossain's *Padmarag* [পদ্মরাগ] (first published in 1924). The chapter queries the representations of space and dwelling in these texts as allegories of the larger socio-spatial community. I show that Kipling presents Indian space as differential, and the Indians made visible in his narrative either work for the colonial state or are simply social types, constructed with a keen ethnographic eye. The spatial horizon of Tagore's novel ranges across the domestic and regional scales and incorporates views of the city and the country that are populated by the Indian elite – anti-colonial and comprador – and the British. His spatial representations make visible peasants, untouchables, and the underclass, gesturing towards Indian society as a spatialized heteroglossia where Hindu parochialism, universal humanism, anti-colonial nationalism, casteism, sexism, and the struggle for women's emancipation co-exist. The text by Hossain, India's pioneering feminist, is largely populated by agential women and set in Tarini Bhavan, a shelter for destitute women. I read Tarini Bhavan as allegorizing the possibility of an egalitarian postcolony where women are at par with men and control their own lives. And finally, in my epilogue, I meditate on the relationship between decolonization and utopia and suggest that decolonization is both hope actualized and a horizon of expectation. It is, as such, both an event and a process. I take up the trajectory of utopianism in postcolonial India suggesting that while it has largely been denuded of content to become an empty formula, the spirit of utopia nevertheless lives on and gives life to transformative politics.

The studies that follow highlight how British and Bengali texts contest ideas of India to articulate what India is and who Indians are; to whom India belongs. *Space, Utopia and Indian Decolonization* illuminates how the literary texts and the literary imagination allowed Indians to reclaim the space of Britain's Indian colony and redefine it as a postcolony. It also shows the texts' re-imagination of the British Indian subjects as a community of putative citizens. This process of contestation – and this is important to bear in mind – is not just, or simply, between colonial and counter-colonial desires but also, more crucially perhaps, between different kinds of counter-colonial desires that gesture towards radically different utopian horizons. To say this another way: the counter-colonial desires point to very different ideas of what India is and could become even as they set down ideas of who is – or is not – an Indian. The contestatory process of literary imagination that I outline in the pages that follow focuses on a specific moment in South Asia's colonial period. But the process as such is by no means restricted only to the colonial period. In fact, similar processes of contestation are in evidence in contemporary India, in this second decade of the twenty-first century, when different literary and cultural texts with their diverse kinds of spatial desires are seeking to fashion the identity of India and Indians once again in distinct and distinctive ways.

Notes

1 For a helpful overview of the category, see Jennifer Wenzel, "Decolonization," in *A Companion to Critical and Cultural Theory*, eds. Imre Szeman, Sarah Blacker, and Justin Sully (Hoboken: Wiley Blackwell, 2017), 449–464.
2 Frantz Fanon, *The Wretched of the Earth*, trans. Constance Farrington (New York: Grove Press, 1963), 37–38.
3 Edward Said, *Culture and Imperialism* (New York: Vintage Books, 1994), 7.
4 I draw on the term "imagined geographies" (sometimes referred to as "imaginative geographies") from Edward Said. For an elaboration of it, see Edward Said, *Orientalism* (New York: Vintage Books, 1979), 49–72. My understanding of the "one and unequal colonial literary world-system" develops out of theorizations of World Literature by Franco Moretti, Pascale Casanova, and the Warwick Research Collective. See Franco Moretti, "Conjectures on World Literature," *New Left Review* 1 (2000), 54–68; Pasacale Casanova, *The World Republic of Letters*, trans. M. B. DeBevoise (Cambridge: Harvard University Press, 2004); and Warwick Research Collective, *Combined and Uneven Development: Towards a New Theory of World-Literature* (Liverpool: Liverpool University Press, 2015).
5 My formulation is indebted to Ernst Bloch, *The Principle of Hope, Volumes 1–3*, trans. Neville Plaice, Stephen Plaice, and Paul Knight (Cambridge: MIT Press, 1986); Raymond Williams, *Marxism and Literature* (Oxford: Oxford University Press, 1977); Fredric Jameson, *The Political Unconscious: Narrative as a Socially Symbolic Act* (Ithaca: Cornell University Press, 1982); and Louis Marin, *Utopics: The Semiological Play of Textual Spaces*, trans. Robert A. Vollrath (Amherst: Humanity Books, 1984).
6 I borrow the term "postcolony" from Achille Mbembe's seminal work. But my inflection of the term differs significantly from Mbembe as it does not suggest a decolonized or postcolonial space but rather a pre-figuration of the decolonized version of the colony. See Achille Mbembe, *On the Postcolony* (Berkeley: University of California Press, 2001).
7 For an elaboration of "structure of feeling," see Williams, *Marxism and Literature*, 128–135.
8 Partha Chatterjee, *Nationalist Thought and the Colonial World: A Derivative Discourse?* (Minneapolis, MN: University of Minnesota Press, 1986), 51.
9 One of the earliest – and indeed, classic – studies on the relationship between social space, literature is Raymond Williams, *The Country and the City* (New York: Oxford University Press, 1973). Williams draws out the role of the capitalist mode of production in transforming space in Britain and the world through colonial expansion and extraction. Edward Said's *Culture and Imperialism* furthers Williams's critical agenda and outlines with greater clarity a reading method that connects narrative to social space as well as empire to secular criticism. While Said's study makes the connection between imperialism and culture more explicit, unlike Williams however, it is incurious about the role of capitalism in the production of space and obscures its defining role in global imperialism. It must also be noted that Said's comment in *Culture and Imperialism* that imperialism is peripheral to the main idea of *The Country and the City* is, frankly, mistaken and misleading.

10 Said, *Culture and Imperialism*, 52.
11 Ibid., 58, italics in original.
12 See, for instance, Alison Blunt and Gillian Rose, eds., *Writing Women and Space: Colonial and Postcolonial Geographies* (New York: Guilford Press, 1994); James Graham, *Land and Nationalism in Fictions from Southern Africa* (New York: Routledge, 2009); Sara Upstone, *Spatial Politics in the Postcolonial Novel* (Farnham: Ashgate, 2009); Alan G. Johnson, *Out of Bounds: Anglo-Indian Literature and the Geography of Displacement* (Honolulu: University of Hawai'i Press, 2011); Robert Tally Jr., ed., *Geocritical Explorations: Space, Place and Mapping in Literary and Cultural Studies* (New York: Palgrave Macmillan, 2011); Andrew Teverson and Sara Upstone, eds., *Postcolonial Spaces: The Politics of Place in Contemporary Culture* (Basingstoke: Palgrave Macmillan, 2011); Rashmi Varma, *The Postcolonial City and its Subjects: London, Nairobi, Bombay* (New York: Routledge, 2012); Stanka Radović, *Locating the Destitute: Space and Identity in Caribbean Fiction* (Charlottesville: University of Virginia Press, 2014); John Thieme, *Postcolonial Literary Geographies: Out of Place* (London: Palgrave Macmillan, 2016); and Robert Zacharias, "Space and the Postcolonial Novel," in *The Cambridge Companion to the Postcolonial Novel*, ed. Ato Quayson (New York: Cambridge University Press, 2016), 208–229.
13 Karl Marx, *Capital: A Critique of Political Economy. Volume I, The Process of Capitalist Production*, trans. Samuel Moore and Edward Aveling (London: Lawrence & Wishart Ltd., 1967), 173.
14 Neil Smith, *Uneven Development: Nature, Capital, and the Production of Space* (Athens: University of Georgia Press, 1990), 107–108.
15 See Henri Lefebvre, *The Production of Space*, trans. Donald Nicholson-Smith (Oxford: Blackwell, 1991).
16 Ibid., 229–291.
17 Ibid., 285.
18 See, for instance, Samir Amin, *Unequal Development: An Essay on the Social Formations of Peripheral Capitalism*, trans. Brian Pearce (New York: Monthly Review Press, 1976); Immanuel M. Wallerstein, *The Capitalist World-Economy: Essays* (Cambridge: Cambridge University Press, 1979); Fredric Jameson, *A Singular Modernity: Essay on the Ontology of the Present* (London: Verso, 2002); and E. San Juan Jr., "Postcolonialism and the Problematic of Uneven Development," in *Marxism, Modernity and Postcolonial Studies*, eds. Crystal Bartolovich and Neil Lazarus (Cambridge: Cambridge University Press, 2002), 221–239. For a recent attempt to understand the literary registration of combined and uneven development, see Warwick Research Collective, *Combined and Uneven Development*.
19 Martin Heidegger, "The Age of the World Picture," in *The Question Concerning Technology and Other Essays*, trans. William Lovitt (New York: Harper Perennial, 1977), 115–154.
20 Smith, *Uneven Development*, 116.
21 Thongchai Winichakul, *Siam Mapped: A History of the Geo-Body of a Nation* (Honolulu: University of Hawaii Press, 1994), 79.
22 Ibid., 75.
23 The evolution of the border-concept in the South Asian context remains under-studied. One of the earliest attempts to engage with this question is Ainslee Embree, "Frontiers into Boundaries: From the Traditional to the Modern State," in *Realm and Region in Traditional India, Volume 2*, ed. R. Fox (Durham, NC: Duke University Press, 1977), 255–270. For a critique of Embree, see Ian Barrow, *Making History, Drawing Territory: British Mapping in India c.1756–1905* (New Delhi: Oxford University Press, 2003). The cartographic anxiety regarding borders in postcolonial South Asia is examined by Shankaran Krishna, "Cartographic Anxiety: Mapping the Body Politic in India," in *Challenging Boundaries: Global Flows, Territorial Identities*, eds. M. Shapiro and H. Alker (Minneapolis, MN: University of Minnesota Press, 1996),

193–214. For an engagement with territoriality in South Asia, and the histories of the Durand, McMahon, and Radcliff lines, see A. Mishra, "Boundaries and Territoriality in South Asia: From Historical Comparisons to Theoretical Considerations," *International Studies* 45 (2008), 105–132.

24 Studies on nationalism have typically been incurious of space. This is perhaps most evident in Benedict Anderson's seminal study on nationalism, *Imagined Communities*. Anderson addresses his lack of engagement with space in the second, revised edition of the text by inserting a chapter called "Census, Map, Museum." In a self-critical comment about the earlier edition of his work he writes about realizing "what I had believed to be a significantly new contribution to thinking about nationalism – changing apprehensions of time – patently lacked its necessary coordinate: changing apprehensions of space." See Benedict Anderson, *Imagined Communities: Reflections on the Origin and Spread of Nationalism* (London: Verso, 1991), xiii–xiv. This is also true of scholars such as Partha Chatterjee and Dipesh Chakrabarty who despite giving their texts titles that evoke spatial registers, nevertheless remain incurious about space. See Partha Chatterjee, *The Nation and its Fragments: Colonial and Postcolonial Histories* (Princeton, NJ: Princeton University Press, 1993); Chatterjee, *Nationalist Thought and the Colonial World*; and Dipesh Chakrabarty, *Provincializing Europe: Postcolonial Thought and Historical Difference* (Princeton, NJ: Princeton University Press, 2000), 149–179. This lacuna is substantially addressed in the path-breaking work of Manu Goswami which details Indian anti-colonial nationalism as a process of transforming the colonial state-space of British India to the Indian nation-space. While Goswami opens her study noting that "central to the project of nationalism is making the nation appear natural," she does not, however, engage with the domain of the literary in shaping the imagination of the Indian nation-space and naturalizing it. Unsurprisingly, her otherwise compelling work does not engage with the question of ideology in the making of the Indian nation. See Manu Goswami, *Producing India: From Colonial Economy to National Space* (Chicago: University of Chicago Press, 2004), 1. Of these theorists, the most astute thinker of the literary remains Anderson who posits "print capitalism," and especially the novel and the newspaper as harbingers of national consciousness. For a study that discusses the influence of colonial geographical knowledge systems on hierarchical conceptions of space among Indian elite classes, see Subho Basu, "The Dialectics of Resistance: Colonial Geography, Bengali Literati and the Racial Mapping of Indian Identity," *Modern Asian Studies* 44, no. 1 (2010): 53–79.

25 In the works of Chatterjee, Chakrabarty, and Goswami, for instance, the Indian nation is implicitly presumed to be the necessary outcome of decolonization and, more problematically, presumed to be a singular entity. This is particularly curious in the case of Chatterjee and Chakrabarty who were pre-eminent members of the Subaltern Studies Collective that, at least in its originary form, critiqued the idea of a singular Indian nation. See Chatterjee, *The Nation and its Fragments*; Chatterjee, *Nationalist Thought*; Chakrabarty, *Provincializing Europe*; and Goswami, *Producing India*. For a delineation of the changing character of Subaltern Studies, see the chapter "The Decline of the Subaltern in *Subaltern Studies*" in Sumit Sarkar, *Writing Social History* (New Delhi: Oxford University Press, 1997), 82–108 and Timothy Brennan, "Subaltern Stakes," *New Left Review*, 89 (September October 2014): 67–87.

26 Sugata Bose, "Nation as Mother: Representations and Contestations of 'India' in Bengali Literature and Culture," in *Nationalism, Democracy, and Development: State and Politics in India*, eds. Sugata Bose and Ayesha Jalal (New Delhi: Oxford University Press, 1997), 70.

27 Neil Lazarus, *The Postcolonial Unconscious* (Cambridge: Cambridge University Press, 2011), 65.

28 Chakrabarty, *Provincializing Europe*, 149.

29 Neil Lazarus, *Nationalism and Cultural Practice in the Postcolonial World* (Cambridge: Cambridge University Press, 1999), 68–143. Also see Lazarus, *The Postcolonial Unconscious*, 21–88.

30 Ruth Levitas, *The Concept of Utopia* (Oxford: Peter Lang, 1990), 6–8.
31 Ibid., 8.
32 Ernst Bloch, *The Principle of Hope, Volume 1*. See, in particular, Part Two (Foundation) which is subtitled "anticipatory consciousness."
33 In addition to the studies by Jameson, Wegner, and Tally cited earlier I would add Ruth Levitas, *Utopia as Method: The Imaginary Reconstitution of Society* (New York: Palgrave Macmillan, 2013); Lyman T Sargent, "The Three Faces of Utopianism Revisited," *Utopian Studies* 5, no. 1 (1994): 1–37; and Fredric Jameson, "Utopia as Method, or the Uses of the Future," in *Utopia/Dystopia: Conditions of Historical Possibility*, eds. M. D. Gordin, H. Tilley, and G. Prakash (Princeton, NJ: Princeton University Press, 2010), 21–44.
34 Walter Benjamin, *The Origin of German Tragic Drama*, trans. John Osborne (London: Verso, 1998), 34.
35 See Theodore Adorno, *Negative Dialectics*, trans. E. B. Ashton (London: Routledge & Kegan Paul Ltd., 1973). For an elaboration on constellation in Adorno and Benjamin, see Fredric Jameson, *Late Marxism: Adorno, or the Persistence of the Dialectic* (London: Verso, 1990), 49 –58.
36 Said, *Culture and Imperialism*, 51.
37 Ibid.
38 Ibid.
39 Jacqueline Dutton and Lyman Tower Sargent, "Introduction: Utopias from Other Cultural Traditions," *Utopian Studies* 24, no. 1 (2013): 2–5.
40 Some of the studies I have in mind include Fredric Jameson, *Archaeologies of the Future: The Desire Called Utopia and Other Science Fictions* (London: Verso, 2005); Philip E. Wegner, *Imaginary Communities: Utopia, Nation, and the Spatial Histories of Modernity* (Berkeley: University of California Press, 2002); Robert Tally, *Utopia in the Age of Globalization: Space, Representation, and the World-System* (Basingstoke: Palgrave Macmillan, 2013). The limited critical corpus that focuses on texts from the "non-West" include Nicholas Brown, *Utopian Generations: The Political Horizon of Twentieth-Century Literature* (Princeton, NJ: Princeton University Press, 2005); Lyman Tower Sargent, "Colonial and Postcolonial Utopias," in *The Cambridge Companion to Utopian Literature*, ed. Gregory Claeys (Cambridge: Cambridge University Press, 2010), 200–222; Jacqueline Dutton, " 'Non-western' utopian traditions," in *The Cambridge Companion to Utopian Literature*, ed. Gregory Claeys (Cambridge: Cambridge University Press, 2010), 223–258; Anupama Mohan, *Utopia and the Village in South Asian Literatures* (Basingstoke: Palgrave Macmillan, 2012); and some of the essays gathered in the collection *The Politics of the (Im)Possible*. See for instance: Barnita Bagchi, "Ladylands and Sacrificial Holes: Utopias and Dystopias in Rokeya Sakhawat Hossain's Writings," in *The Politics of the (Im)Possible: Utopia and Dystopia Reconsidered*, ed. Barnita Bagchi (New Delhi: Sage Publications, 2012), 166–178; Subhoranjan Dasgupta, "Dystopia, Utopia, and Akhtaruzzaman Elias's Novel *Khowabnama*," in *The Politics of the (Im)Possible: Utopia and Dystopia Reconsidered*, ed. Barnita Bagchi (New Delhi: Sage Publications, 2012), 80–94; Modhumita Roy, "Utopia in the Subjunctive Mood: Bessie Head's *When Rain Clouds Gather*," in *The Politics of the (Im)Possible: Utopia and Dystopia Reconsidered*, ed. Barnita Bagchi (New Delhi: Sage Publications, 2012), 179–193; and Samita Sen, "A Parliament of Women: Dystopia in Nineteenth-century Bengali Imagination," in *The Politics of the (Im)Possible: Utopia and Dystopia Reconsidered*, ed. Barnita Bagchi (New Delhi: Sage Publications, 2012), 123–146.
41 For a reading of Bengali culture in a global context, see Andrew Sartori, *Bengal in Global Concept History: Culturalism in the Age of Capital* (Chicago: University of Chicago Press, 2008). For some attempts to think about Victorian literature in a global frame, see Tanya Agathocleous, "Imperial, Anglophone, Geopolitical, Worldly: Evaluating the 'Global' in Victorian Studies," *Victorian Literature and Culture* 43, no. 3 (2015): 651–658; Brenda Assael, "Beyond Empire: Globalizing the Victorians,"

Victorian Literature and Culture 43, no. 3 (2015): 643–650; Priya Joshi, "Globalizing Victorian Studies," *The Yearbook of English Studies* 41, no. 2 (2011): 20–40; Pablo Mukherjee, "Introduction: Victorian World Literatures," *The Yearbook of English Studies* 41, no. 2 (2011): 1–19.

42 Chakrabarty, *Provincializing Europe*, 28. For a call to rethink the discipline of Comparative Literature that positions the non-west as active producers of literature and culture, see Gayatri Spivak, *Death of a Discipline* (New York: Columbia University Press, 2003).

43 Ngũgĩ wa Thiong'o, *Globalectics: Theory and the Politics of Knowing* (New York: Columbia University Press, 2012), 8.

44 For a discussion of the plural literary field of South Asia and the category of "Indian Literature," see chapter 7 of Aijaz Ahmad, *In Theory: Classes, Nations, Literatures* (London: Verso, 1992). Also see, Francesca Orsini, "India in the Mirror of World Fiction," *New Left Review* 13 (2002): 75–88 and Amiya Dev, "Comparative Literature in India," CLCWeb: *Comparative Literature and Culture* 2, no. 4 (2000). https://doi.org/10.7771/1481-4374.1093. For a discussion of the Punjabi, see Farina Mir, *The Social Space of Language: Vernacular Culture in British Colonial Punjab* (Berkeley: University of California Press, 2010).

45 See chapter 1 of Nikhil Govind, *Between Love and Freedom: The Revolutionary in the Hindi Novel* (New Delhi: Routledge, 2014). Also see T. W. Clark, ed., *The Novel in India: Its Birth and Development* (Berkeley: University of California Press, 1970).

46 For an illuminating discussion of the Bengali literary canon that situates it in relation to the erasures and appropriations of postcolonial literary studies, see Auritro Majumder, "Can Bengali Literature be Postcolonial?," *Comparative Literature Studies* 53, no. 2 (2016): 417–425.

47 The corpus of book-length studies in the English language that engage exclusively with non-Anglophone South Asian literary texts from the colonial and postcolonial eras, or even in conjunction with Anglophone texts, remain exceedingly small. These include, but are not limited to, T. W. Clark, ed., *The Novel in India*; Meenakshi Mukherjee, *Realism and Reality: The Novel and Society in India* (New Delhi: Oxford University Press, 1985); Frances W. Pritchett, *Nets of Awareness: Urdu Poetry and Its Critics* (Berkeley: University of California Press, 1994); Sudipta Kaviraj, *The Unhappy Consciousness: Bankimchandra Chattopadhyay and the Formation of Nationalist Discourse in India* (New Delhi: Oxford University Press, 1995); Sangeeta Ray, *En-Gendering India: Woman and Nation in Colonial and Postcolonial Narratives* (Durham, NC: Duke University Press, 2000); Priyamvada Gopal, *Literary Radicalism in India: Gender, Nation and the Transition to Independence* (London: Routledge, 2005); Tridip Suhrud, *Writing Life: Three Gujarati Thinkers* (New Delhi: Orient Blackswan, 2009); Ulka Anjaria, *Realism in the Twentieth-Century Indian Novel: Colonial Difference and Literary Form* (New York: Cambridge University Press, 2012); Subramanian Shankar, *Flesh and Fish Blood: Postcolonialism, Translation, and the Vernacular* (Berkeley: University of California Press, 2012); Mohan, *Utopia and the Village*; Toral Jatin Gajarawala, *Untouchable Fictions: Literary Realism and the Crisis of Caste* (New York: Fordham University Press, 2013); Govind, *Between Love and Freedom*; Rosinka Chaudhuri, *The Literary Thing: History, Poetry, and the Making of a Modern Cultural Sphere* (New Delhi: Oxford University Press, 2014); Sudipta Kaviraj, *The Invention of Private Life: Literature and Ideas* (New York: Columbia University Press, 2015); Udaya Kumar, *Writing the First Person: Literature, History, and Autobiography in Modern Kerala* (Ranikhet: Permanent Black, 2016); Kiran Keshavamurthy, *Beyond Desire: Sexuality in the Modern Tamil Novel* (New Delhi: Oxford University Press, 2016); Krupa Shandilya, *Intimate Relations: Social Reform and the Late Nineteenth Century South Asian Novel* (Evanston: Northwestern University Press, 2017); Amit R. Baishya, *Contemporary Literature from Northeast India: Deathworlds, Terror and Survival* (London: Routledge, 2018); and G. S. Sahota, *Late Colonial Sublime: Neo-Epics and the End of Romanticism* (Evanston: Northwestern University Press, 2018).

48 Rabindranath Tagore, "chitto jetha bhoyshunyo," [চিত্ত যেথা ভয়শূন্য] in *Noibedyo* [নৈবেদ্য] (Calcutta: Vishwabharati, 1901), 83. The translation is mine. The original lines are:

> চিত্ত যেথা ভয়শূন্য, উচ্চ যেথা শির,
> জ্ঞান যেথা মুক্ত, যেথা গৃহের প্রাচীর
> আপন প্রাঙ্গণতলে দিবসশর্বরী
> বসুধারে রাখে নাই খণ্ড ক্ষুদ্র করি,
> যেথা বাক্য হৃদয়ের উৎসমুখ হতে
> উচ্ছ্বসিয়া উঠে, যেথা নির্বারিত স্রোতে
> দেশে দেশে দিশে দিশে কর্মধারা ধায়
> অজস্র সহস্রবিধ চরিতার্থতায় –
> যেথা তুচ্ছ আচারের মরুবালুরাশি
> বিচারের স্রোতঃপথ ফেলে নাই গ্রাসি,
> পৌরুষেরে করেনি শতধা; নিত্য যেথা
> তুমি সর্ব কর্ম চিন্তা আনন্দের নেতা –
> নিজ হস্তে নির্দয় আঘাত করি, পিতঃ,
> ভারতেরে সেই স্বর্গে করো জাগরিত |

49 Vinayak Damodar Savarkar, *Hindutva: Who is a Hindu?* (Bombay: Veer Savarkar Prakashan, 1969), 111.
50 Ibid., 113.
51 John Strachey, *India* (London: Kegan Paul, Trench & Co., 1888), 2.
52 Ibid., 5.

1 Of good and evil

The anxiety of utopianism

Romance, Northrop Frye tells us, is "the structural core of all fiction: being directly descended from folktale, it brings us closer than any other aspect of literature to the sense of fiction, considered as a whole, as the epic of the creature, man's vision of his own life as a quest."[1] Developing Frye's theorizations further, Jameson writes that:

> romance is . . . a wish-fulfilment or Utopian fantasy which aims at the transfiguration of the world of everyday life in such a way as to restore the conditions of some lost Eden, or to anticipate a future realm from which the old mortality and imperfections would have been effaced. Romance . . . does not involve the substitution of some more ideal realm for ordinary reality. . . . But rather a process of transforming ordinary reality.[2]

He goes on to note that romances are structured as struggles between higher and lower realms, between heaven and hell, or the angelic and the demonic or diabolic – in sum, a clash between good and evil.[3] The utopianism of the romance genre expresses itself, in particular, in the narrative denouements. They provide, Jameson suggests, an

> imaginary "solution" to . . . [a] real contradiction, a symbolic answer to the perplexing question of how my enemy can be thought of as being evil (that is, as other than myself and marked by some absolute difference), when what is responsible for his being so characterized is quite simply the identity of his own conduct with mine.[4]

While the utopian lineaments of the romance genre have been drawn out with relation to canonical texts of the "West," critics have rarely, if at all, taken up their colonial or counter-colonial ramifications.[5] In this chapter, I engage with romances from colonial India, especially those from the decades following the 1857 Uprising, to illuminate their colonial and counter-colonial spatial desires. I seek to *globalize* the scholarly insights on the romance genre that have emerged from Euro-American texts and shed light on a different articulation of utopianism in the majority world of the non-west.

In what follows, I constellate and read two romances from colonial India from the decades following the 1857 Uprising: Philip Meadows Taylor's "mutiny novel" *Seeta* from 1872, and Bankim Chandra Chatterji's Bengali novel *Anandamath* [আনন্দমঠ] published exactly a decade later, in 1882.[6] My reading of these texts illuminates the nexus between imagined geographies, spatial utopianism and the romance form in the colonial context. I show that these romances draw on the antinomy of good and evil, so central to the genre, by encoding their narrative spaces with notions of good and evil. Read together, the colonial and counter-colonial spatial desires of these texts perform a kind of Gestalt alternation between ideological codes in relation to the territorial expanse of British India, signifying the same space, respectively, as evil or good. In other words, they posit British India as a colony or a postcolony. The colony and postcolony are social spaces and, as Lefebvre reminds us, necessarily social products: we cannot really think of the spatial without human intervention and, therefore, human subjects.[7] The spatialization of the categories of good and evil, thus, has a necessary corollary: good and evil subjects. This imbrication of space and subjectivity allows these texts to engage with, and attempt to settle, questions of belonging. The socio-spatial imaginaries of the romances posit who may and may not inhabit those spaces.

But all utopianisms – and this includes the utopianism of the romance genre as well – are also, and always, haunted by the possibility of the failure to actualize their utopian longings. This is because the principle of hope that undergirds these utopias is never confident of its actualization and is always open to disappointment. Bloch glosses this fundamental contradiction of utopian longing noting that this hope is "the opposite of security. It is the opposite of naive optimism. The category of danger is always within it. This hope is not confidence. . . . If it could not be disappointed, it would not be hope."[8] The sense of its possible impossibility is a critical condition of the utopian principle. In the context of de/colonization of South Asia, the awareness of the possibility of transforming Indian space into a post/colony always, and necessarily, goes hand in hand with the awareness of that these efforts may, in fact, end in failure. There is thus a fundamental, and indeed dialectical, relationship between utopia and anxiety. Utopias, thus, are always anxious. In the case of romances, this translates into specific kinds of narrative anxiety regarding their spaces and subjects they fashion. Consequently, these romances formulate devices to ensure narrative closures of specific kinds or devise strategies to expel the evil subject from their narrative worlds. In so doing they partake in structures of socio-spatial wish-fulfilment and catharsis that gesture towards the ritual function of literature in capitalist modernity.[9] And just as crucially, such imaginary resolutions of narrative crises in the symbolic realm mediate and refract very real crises from the world beyond.[10]

The anxiety of colonialism: Philip Meadows Taylor's *Seeta*

In 1857, much of India was convulsed by a bloody uprising against the British. Though initially a mutiny of the disgruntled sepoys of the Bengal Army, the insurrection soon spread to many parts of India, where the mutineers were joined by local rulers such as Nana Sahib and Laxmi Bai who were disinherited by the British. The ensuing hostilities turned the 1857 Uprising (which the British termed "Sepoy Mutiny" and the Indian nationalists, subsequently, called "The First War of Independence") into the most forceful and significant challenge to British rule in India or anywhere in the Empire at that time.[11] As the British colonizers and the Indians fought each other, Delhi, Kanpur, Lucknow, and Meerut witnessed some of the most sustained and violent clashes that turned these cities into sites of gruesome massacres by the Indians and even more horrific reprisals by the British. Though the Indians were initially successful, by the end of 1858 the British managed to re-establish their power in the Indian subcontinent. The 1857 Uprising hastened the end of the East India Company rule in India and the country came under the direct rule of the British crown. However, the impact of the Uprising went well beyond changes in the mode of colonial governance and the reorganization of the armed forces in British India. It marked a decisive shift in the colonial ordering of power. The Rebellion and its aftermath heralded the mobilization of all the mechanisms of the colonial apparatus, from repressive to ideological, in addition to the deployment of "revitalized programs for greater hegemonic control – economic, political, cultural, and epistemic – over the Indian population."[12]

The 1857 Uprising left a significant mark on the visual and literary texts that emerged in Britain and British India in its aftermath. Britain was flooded with paintings, illustrations, and cartoons; photographs of the sites connected to the Uprising by John Murray and Felice Beato were displayed in Britain before being published as books and woodcuts.[13] These images depicted the aftermath of the war and the British victory over the Indian rebels. The Uprising also gave rise to a huge number of eyewitness accounts, journal articles, histories, novels, poems, and plays about India in Britain. Among the earliest of the eyewitness accounts were *Lady's Diary of the Siege of Lucknow Written for Perusal of Friends at Home* (1858) by a "Mrs James P. Harris," Robert Gibney's *My Escape from the Mutinies of Oudh* (1858), and Mowbray Thompson's *Story of Cawnpore* (1859).[14] Histories of the Uprising such as Charles Ball's *History of the Indian Mutiny* (1858) and George Trevelyan's *Cawnpore* (1866) soon followed.[15] Most significantly for my purpose here, the Uprising structured the emergence of a specific kind of sensational fiction that came to be known as "mutiny novels." Structured as romances, these texts were immensely popular on publication and became influential in moulding British public opinion against the Uprising and the Indians mutineers. They portrayed the Indians as irredeemably evil and sexually depraved, and the colony as "mired in changeless patterns of superstition and violence which can be dominated but not necessarily altered for the better."[16] These accounts produced an image of India and Indians as hostile and in need of pacification thus laying out the colonizers' horizon of political (and military) action: of turning the dystopian space of India into a colonial utopia, that is, into a colony. In other words, they present ideas of good and evil subjects (the *good* colonial administrator; the *good* native; the *evil* mutineer) while also articulating ideas of good and evil versions of British Indian space, that is, British India as mutinous (bad) and as a colony (good).

Philip Meadows Taylor's *Seeta* is a "mutiny novel" but one that is atypical in that it folds the events of the 1857 Uprising into the love story of the novel's eponymous heroine and the colonial official Cyril Brandon. Furthermore, it is also one of the only colonial texts to feature an inter-racial marriage, between the Hindu widow Seeta and Brandon. As the novel opens, readers encounter Azrael Pande the "villain" of the novel before being introduced to Seeta, the wife of a prosperous businessman of Noorpoor, Huree Das. Soon after, in a robbery engineered by Huree Das's cousin Ram Das, Pande murders Seeta's husband and child. The colonial authorities eventually arrest Pande and the case is tried by the Acting Deputy Commissioner of Noorpoor, Cyril Brandon who is impressed by Seeta's poise in court as well as her calm beauty. Seeta's testimony gets Pande convicted but he manages to escape, vowing vengeance on her. Subsequently, on learning that Pande and his men plan to attack the house where Seeta is staying and abduct her, Brandon rushes there and saves Seeta but is injured in the fighting. As Seeta nurses him back to health, they fall in love with each other inaugurating the inter-racial romantic plot of the novel that makes this "mutiny novel" quite singular. In time, Brandon seeks to marry Seeta and the prospect of an inter-racial marriage makes some sections of the Indians as well as the British quite unhappy. Meanwhile, Pande stirs up trouble among the Indians soldiers ("sepoys") and as trouble breaks out across the Indo-Gangetic plains, the British at Noorpoor prepare themselves for a rebellion. The 1857 Uprising breaks out across north India, which is eventually crushed by the British who recapture Delhi and re-establish their hold over South Asia. Pande – depicted in the novel as a relative of Mangal Pande, who supposedly started the revolt of the Bengal Army regiment in Barrackpore – escapes from the fighting and goes into hiding with the rebels. As the English celebrate their victory, Pande attacks the house where the festivities are in progress. In the fighting, Seeta dies protecting Brandon who, now freed from his marital obligation, eventually marries Grace Mostyn and returns to England.

The narrative of *Seeta* is unexceptional and lays out in the most formulaic of ways what Frye and Jameson note is the primary function of the romance genre: the delineation and dramatization of a clash between the categories of good and evil. The narrative goes to great lengths to draw its characters out in black and white and it is unsurprising that, given its colonial provenance, Brandon is presented as an embodiment of good. It signals his goodness in myriad ways but most evidently in the description of Brandon's physical features. We are told that he was

> tall, extremely well made, and graceful in his movements. His features were *not perhaps strictly regular*, yet he was, no doubt, *very handsome*. What you saw at once were a broad, white forehead, over which his glossy brown hair clustered in heavy waves, strong eyebrows, with very clear, earnest blue eyes, shaded by long dark eyelashes, a pleasant mouth, a bright complexion, very white teeth, with a full chin, in which, and in the face generally, there was much character and determination. He wore both moustache and beard, of a rich brown, soft and curly; and his complexion, always pale in England, was now ruddy and healthy.[17]

The description leaves no doubt in the reader's mind about his good looks and underlines it further by suggesting that he was "very handsome." It enumerates his features in a blazon-like manner as we are told of his "broad, white forehead," "glossy brown hair," the "strong eyebrows," "clear, earnest blue eyes," "bright complexion," "very white teeth," and so on. The description of Brandon's looks is complemented by a detailed delineation of his training and abilities, which make him an able colonial administrator. He had, we are told,

> studied hard at Haileybury; carried off repeated prizes in Sanscrit [sic], of which he was very fond, and in Persian, law, and other branches of the college course; and his perfect acquaintance with French and Italian, gained as a boy in the frequent residence . . . at Paris, Florence, and Rome, gave him, a great advantage over his fellow-students in regard to modern languages. He was deficient, no doubt, in Greek and Latin; but had gained enough of both to preserve him from actual failure. He was an excellent draughtsman and artist . . . [and could] both . . . play and . . . sing.[18]

The description here presents Brandon as a conscientious student and a budding "orientalist," the term understood as someone who studies the orient.[19] He was well versed in Sanskrit and Persian, the pre-colonial languages of South Asia, as well as in law, thus making him a version of William Jones, and a literary figure who anticipates Colonel Creighton of Kipling's *Kim*. But besides being well versed in oriental languages and being a graduate of Haileybury, Brandon is also presented as someone who was cultured and well traveled in the European continent. Not only had he made "perfect acquaintance with French and Italian," he also had passing knowledge of the European classical languages Greek and Latin; besides being trained in music, he was "an excellent draughtsman and artist" too. The narrative also stresses his martial abilities for we are told that "there was no better shot, either with gun or rifle, and no bolder rider in the country."[20] In sum, the character of Brandon is presented as a kind of Renaissance man; a *l'uomo universale* who is at ease in the worlds of work as well as of leisure.

In addition to his looks and abilities, the narrative also underlines Brandon's aristocratic lineage. We are told that he was

> descended from an ancient family which was possessed of considerable estates in ______shire; and Hylton Hall, the family seat, a noble specimen of the Tudor style of domestic architecture, was one of the show places of the county. In the reign of Queen Anne, George Hylton, conspicuous for his loyalty and important local services, was raised to the peerage under the title of Baron Hylton. . . . Many thought . . . that the acceptance of a peerage was unnecessary and derogatory, and might have been avoided or declined; but the person on whom it was conferred was an ardent loyalist, and held that compliance with the wish of his Sovereign was his true test of duty.[21]

The passage not only describes the Brandon family seat but also notes the granting of the peerage to Brandon's grandfather, George Hylton by Queen Anne.

Besides stressing Brandon's provenance, it also lays out his grandfather as an "ardent loyalist" whose "true test of duty" was the "compliance with the wish of the Sovereign." This characterization is important not least since Brandon would be key in preserving the rule of law and good government in British India, especially during the 1857 Uprising. And he is also someone who is well liked by people he governs:

> His never-failing affability and good temper, his cheerful patience, his ready accessibility, even to the most humble, his evident desire for the improvement and advancement of all classes, and his efforts . . . to soften down the hard mechanical actions and rules of official life – had won him . . . more than ordinary popularity . . . year by year, the fine qualities of his character had become more and more appreciated. . . . [M]any a dame . . . repeated his name with that of her household gods each night, as she lighted the lamp before the simple shrine of her faith, and taught her children to say it. Many a rude village poet had written ballads, and the minstrels had sung them to his praise, at village festivals.[22]

Brandon here is presented not just as a *good* colonial administrator but as an *ideal* one. Not only does he rule the people well but also does so consensually for he can "soften" the "hard mechanical actions and rules of official life." And even as the narrative draws out the notion of the good colonial administrator, it also posits ideas of the good colonial subjects. Notice how it talks about "many a dame" chanting his name to the gods during evening prayer and village poets and minstrels writing and singing ballads in his praise. These not only point to a kind of colonial utopianism regarding the good ruler and subject but also articulates a specific spatial desire that posits an image of the ideal colony that is ruled with an iron fist in a velvet glove. And where the rulers are accepted, idealized, and praised by their Indian subjects. *Seeta* fulfils Shuchi Kapila's characterization of colonial romances from British India: it encodes "a political fantasy of creating through a process of benevolent rule, native subjects acculturated to European values who welcome colonial rule and ally themselves with British interests."[23]

Brandon, however, is not the only epitome of good in the novel for it also draws out the eponymous heroine as another version of goodness. This is best encapsulated in Brandon's first encounter with Seeta at the Noorpoor courthouse.

> The large dewy eyes were soft and pleading, but not irresolute, and the girl was quite calm. Seeta had dressed her self in a rich silk saree of a green colour, shot with crimson . . . and the end, which she had passed over her head, fell on her right arm and contrasted vividly with *its fair colour* and rounded outline. If her features were *not exactly regular*, they were very sweet and full of expression; her eyes were large and soft, of that clear dark brown which, like a dog's, is always so loving and true. If the mouth were a shade too full for exact symmetry, it was mobile and expressive, and the curves of the upper lip constantly varied. For a native woman, Cyril Brandon had never seen any *one so fair* or of so tender a tone of colour. Such . . . were many of the lovely

> women of Titian's pictures – a rich golden olive, with a bright carnation tint rising under the skin – and Seeta's was like them.[24]

This extract depicts Seeta's physical features in great detail. These descriptions – of her complexion, her eyes, her mouth – parallel the manner in which Brandon is presented to the readers and, as such, gesture towards two articulations of the category of good in the novel. And as in the case of Brandon, it also uses a blazon-like descriptive protocol. Her features, we are told, "were not exactly regular," but she had a "fair colour and rounded outline"; her mouth seemed "too full of exact symmetry" but was nevertheless "mobile and expressive"; and the "curves of her upper lip constantly varied." Seeta, moreover, is described from Brandon's perspective, and her fairness is commented upon more than once. The description notes that her green saree highlights her "fair colour" and then tells readers that Brandon "had never seen any one so fair or of so tender a tone of colour." It then proceeds to render Seeta legible in the European terms suggesting she was like the many "lovely women" of Titian's paintings, now suggesting her complexion was a "rich golden olive" with a hint of red. The passage describes Seeta's eyes twice over. At first, they are described as "large dewy eyes" that appear to be "soft and pleading" and a few sentences later we are again reminded of her eyes, now described as "dark brown" as well as "large and soft." The eyes are now qualified through the image of an animal – like a dog's – and characterized through that simile as "so loving and true." The use of the figure of the dog is actually quite illuminating as it invokes images of tameness and fidelity and anticipates Seeta's characteristics as Brandon's eventual wife. Here again we see a certain structure of colonial utopianism that imagines the ideal Indian subject: the domestic Indian woman who is also, and just as crucially, an object of colonial heterosexual desire "whose docility represents political subservience."[25]

If Seeta and especially Brandon are the epitomes of good in the novel, Azrael Pande is presented as their foil and the embodiment of unalloyed evil. This is powerfully signalled in the naming of the novel's villain. Kapila draws attention to the villain's nomenclature noting that "Azrael is the formidable and ruthless angel of death in Christian, Jewish, and Islamic literature. It is unlikely that Pande, a Brahmin, would have a first name that came from the literature of the Abrahamic religions."[26] While Kapila's observation about the incongruity of the naming is perceptive, I want to suggest that it is precisely the strangeness of this juxtaposition that drives Taylor's characterization of his villain. And this becomes especially moot when considering the context of the novel's reception. The name Azrael would conjure the figure of the angel of death for Taylor's European readers. And at the same time, the name Pande would invoke for them the figure of Mangal Pande, the "sepoy" of the 34th Bengal infantry who supposedly began the rebellion in Barrackpore outside Calcutta that would develop into the "Sepoy Mutiny" or the 1857 Uprising. Azrael, it bears repeating, is depicted as Mangal's uncle in the novel. The name Pande would also recall for the readers the term "pandies," the British term for the mutineers that is derived from Mangal Pande's last name. A household word in Britain and British India in the aftermath of 1857,

Taylor also uses it extensively, and throughout the novel, to refer to the mutineers. The nomenclature, moreover, speaks to the author's stated aim about his characterization of the novel's villain. In the introduction of the novel, Taylor writes that he had

> endeavoured to depict the character of the rebel and treasonable emissaries of the time. Malignant and persistent, they were led on by blind hatred and religious fanaticism, to the instigation and commission of crimes at which humanity shudders.[27]

By naming his villain Azarel Pande, then, Taylor underlines the character as the epitome of malevolence.

Pande is the first character we meet in *Seeta* and Taylor gives us a detailed description of him in a manner that is not unlike his portrayal of Brandon. He describes Pande as a

> tall commanding figure, with a great breadth of chest and strength of limb. . . . *His features were regular*, and would have been handsome but for his mouth, which, when he spoke, disclosed jagged, *irregular* teeth; and, together with some projection of the lower jaw, gave him, when excited, the expression of a wild beast. The coal-black eyes were large, deep set, and fierce in expression. On his forehead, which was high and smooth, a strange knot of wrinkles would often arise suddenly. . . . For a native of India, his complexion might be called fair – a bright olive, which sometimes flushed into dark red. His thick moustache was streaked with grey, and his age might be forty years. . . . By caste he was a Brahmin, and was well read in the sacred Hindoo books; and . . . his recitations to village assemblies during his wanderings were eagerly welcomed and well rewarded.[28]

Taylor's description of Pande draws out his physical features and readers are told that he was a "tall commanding figure" who had a "great breadth of chest" and "strength of limb." And as in the case of Brandon and Seeta, Taylor once again enumerates his character's features in a blazon-like manner. We are told of his "regular" features and one who could have been handsome but for his "disclosed jagged, irregular teeth" that, with his jaw, made him look like a wild beast. The descriptive tenor continues with the delineation of his "coal-black eyes" that were "large, deep set, and fierce in expression."

The words and phrases used to describe Pande closely parallel those used for Brandon and Seeta in the novel. Pande, in fact, is described in a manner *similar* to Brandon and Seeta but his *differences* to these "good" characters are also painstakingly underlined. This is evident, for instance, in the use of the word "regular" to describe the physical features of the characters. If Pande's features were "regular," Brandon's were not "strictly regular" and Seeta's "not exactly regular." Likewise, while Brandon was "very handsome," Pande is described as "would have been handsome." The parallels and contrasts continue in the fairly detailed description of their facial features. Brandon, we are told, had "earnest blue eyes, shaded by long dark eyelashes," while Pande's eyes were "coal-black . . . large, deep set, and fierce in expression" and Seeta's were "dewy," "large and soft" and "clear dark brown." Similarly, the colonial official's "very white teeth" is contrasted to the Indian rebel's "jagged, irregular teeth." Both hero and villain are depicted as fair but not similarly so. Brandon's "bright complexion" that was "always pale" in England was, in India, "ruddy and healthy" whereas Pande "might be called fair – a bright olive, which sometimes flushed into dark red." And here, Seeta is rendered somewhat similar to Pande for her skin is "rich golden olive" with a hint of red. Likewise, if the orientalist is shown as well versed in Sanskrit, Persian, French, and Italian and adept at performance (he could play instruments and sing), the oriental "was well read in the sacred Hindoo books" and adept in another kind of performance: his recitations to villagers were "eagerly welcomed and well rewarded." Seeta is not unlike them as she, we are told, had learned Sanskrit and read some of the Indic classics. But Seeta's similarity and difference to Pande is forcefully underlined through the use of the figure of the animal. As we have seen in the passage above, Seeta is compared to a dog and the description evokes notions of tameness, meekness, fidelity and docility. Pande is also described through the figure of an animal but that serves as a counterpoint to Seeta's. The novel tells us that Pande's irregular teeth and lower jaw make him appear, when agitated, like a wild beast. *Seeta*, thus, sets up not just notions of good and evil but also notions of the good and evil Indian.

The novel's delineation of good and evil subjects is also, and necessarily, tied to its spatial imaginaries. It depicts specific *kinds* of spaces that are invested with a range of qualities, which could be organized under the rubrics of good and evil spaces. This is particularly true in the case of Azrael Pande who is associated with locales which appear to signify the spatialization of evil. This can be seen in the manner in which the novel's opening yokes together the social and the spatial investing the opening with notions of danger, wildness, and indeed, evil. The first few pages of the novel provide an expansive view of space and setting that functions, anachronistically speaking, like a cinematic sweep to establish a spatial idea of India. And the category of "India" is presented as, and through, spatial views that give readers a *feel* of the novel's setting. *Seeta* begins with:

> It was nearly midnight. A slight passing shower had just fallen, and the moon, nearly at its full, shone out with brilliant lustre over a scene at once strikingly beautiful and impressive. At the head of one of the long ravines which descend westwards from the plateau of Central India, was a deep, lonely glen, the upper end of which was closed by one of those abrupt precipices of basalt, which everywhere mark the boundary of the table-land above, and over which a small stream, swelled somewhat beyond its wont by the showers of the rainy season, now poured a considerable body of water with a dull continuous roar. The moonlight fell directly upon the precipice and fall, and lighted them up with vivid effect. As the waters rushed downwards, at first through a narrow gully at the summit of the precipice, they threw up

> ever-changing jets of spray, flashing for a moment like molten silver as they caught the light, and then disappeared.[29]

The passage really sets up a key imagined geography of the novel and presents readers with a very strong sense of place that draws out the physical landscape in great detail. It speaks of the locale as "strikingly beautiful and impressive" and proceeds to give us a detailed portrayal by drawing out the "long ravines," "the deep lonely glen," and "precipices of basalt" in addition to speaking of the swollen stream that the rains have turned into a sort of waterfall. And besides laying stress, overwhelmingly and powerfully, on the visual quality of the scene, the passage also stresses movement. This is done through the repeated emphases on temporal markers: "nearly midnight"; "a . . . shower had just fallen"; the "rainy season." Note also the line that combines the visual with the spatio-temporal: the "ever-changing jets of spray, flashing for a moment like molten silver as they caught the light, and then disappeared." Indeed, the mimetic quality of the extract gestures towards a developing aesthetic of realism and appears to partake in what Jameson calls the aesthetic mode's "affective impulse."[30]

But the novel's opening does more than simply ground the narrative in space and familiarize the reader with the space of its narrative. It actually invests the space with specific qualities. While the description begins with the suggestion that the space was "strikingly beautiful and impressive," it soon takes on a more ominous tenor. As the focalization moves from the generic landscape to the "only level spot in the glen for some miles," readers are told that for locals "the place had an *evil* reputation, and was carefully avoided."[31] By using the word "evil" directly, the novel inserts a sense of the sinister into the space, and therefore begins a process of spatializing evil (and good) that runs through its spatial imaginary. The sense of foreboding is further accentuated in the novel's description of an unusual tree:

> At . . . this level spot, grew a vast peepul tree of enormous size and remarkably picturesque though peculiar character. One half of it still flourished luxuriantly . . . trembling with every breath of wind, sparkled in the strong light. The rest of the tree was dead; and its long, gaunt branches, bleached white by the sun and wind of many centuries, stretched upwards to the precipice, and partly over the pool, in weird, ghastly forms.[32]

The depiction of the tree – it is both dead and alive – is significant and functions as a kind of dialectical image that shows life and death locked together and at a standstill in a kind of spatio-temporal embrace that points to a spatial contradiction. This image of contradiction, concretized in and through the image of the tree, seems to suggest it as an allegory of the larger space of British India. India, as the novel will underscore for its readers throughout, is like the tree insofar as it brings together not just life and death but also the visible with the invisible. It is a space that collapses the known and knowable with the unknown and unknowable, and indeed, good with evil. The image of the tree, in effect, sums up very powerfully the larger problematic of the novel: the interaction and interpenetration of the active and the passive, life and death, and of course, good and evil. And before continuing with my discussion I would also like to draw attention to, if only in passing, the remarkable similarity between the description of the dead part of the tree – the "long, gaunt branches bleached white by the sun" that "stretched upwards to the precipice" – and the iconic description of the Marabar Hills that would be composed about half a century later by E. M. Forster: "Only in the south, where a group of fists and fingers are thrust up through the soil, is the endless expanse interrupted. These fists and fingers of the Marabar Hills, containing the extraordinary caves."[33]

The novel moves from the tree to the "wide square hall, at the end of which, in an apse or recess, was a large statue of Boodh [the Buddha] in a sitting posture . . . [that] looked out benignly into the beautiful glen, and over the sparkling waters of the stream."[34] The space with the image of the Buddha serves as the hideout of a figure that the novel presents as the exact opposite of that embodiment of compassion and goodness, that is, Azrael Pande. The cave temple, which appears to be a space of sanctity becomes its exact opposite and site of wickedness. This is signalled, once again, through the use of the word "evil": it "was not without cause that this place had earned an *evil* reputation by night as well as by day."[35] This is further underscored by investing the space with danger and lawlessness for readers are told that at the back of this cave temple was now a blazing fire around which "were seated a company of about twenty men, whose savage appearance betokened a hard, lawless life" and as leader was none other than Azrael Pande.[36] This strategy of yoking together the Buddha (representing goodness and life) and Azrael Pande (signifying evil and death) continues the novel's investment in highlighting the contradictory nature of space as we have seen in the case of the tree. These images of spatial contradiction – the juxtaposition of the Buddha and Azrael Pande; the tree that is both dead and alive – seem to suggest that appearances in these spaces (and, by extension, in India) could be, and indeed are, deceptive. It also suggests that these spaces, and India, are fundamentally unknowable, deceptive, and indecipherable that can never be fully known. In effect, these images gesture towards a crisis of epistemology that underwrites colonial spatial desire and points to another register of anxiety.

Seeta contrasts the lawless and evil space of the cave temple with that of the law-abiding and good space of the village of Gokulpoor. It is not insignificant that the house of the goldsmith Huree Das, like the wild landscape that opens the novel, is also made of basalt. The houses of the village, however, are built of "neatly squared" basalt stones as well as red brick thereby pointing to the stones not only being marked with human labour but also suggesting an altering and taming of the wildness of the earlier landscape.[37] The village is not just an image of domesticated space but also one of spatial domesticity. It comprises two "large stone houses" around which "clustered a number of more unpretending dwellings inhabited by farmers, and the artizans [sic] necessary to so small a community. A humble temple . . . was the only religious edifice."[38] The description of the life at the goldsmith's house underlines the domesticity and sociality of Gokulpoor's existence with Huree Das shown as a paternal figure to his workers.

> The master rose, and went into an outer verandah, where several of his apprentices were still at work on some silver ornaments which had been commissioned. The execution did not appear to be satisfactory to the master, for he . . . took up the work which had been in progress, and began fashioning a stout silver wire into an intricate pattern, for an anklet, with great dexterity, while the apprentice looked on in amazement at the skill being displayed.[39]

Readers are presented here with the image of a master craftsman and caring master who not only makes his apprentices work for him but also trains them in the craft. It portrays the everyday life of an unremarkable location in British India and underlines the social processes that shape and reshape that space. Here too a structure of colonial spatial desire is noticeable for this image of spatial domesticity, where people engage in quiet industry, appears to be a kind of ideal space inhabited by ideal subjects.

The idealization of Gokulpoor parallels the novel's idealization of various colonial spaces. The structure of spatial desire is noticeable in the description of the English magistrate's "white tent" that Seeta sees coming up from her window:

> she saw . . . a white tent rising slowly under the trees, and knew that the time of her examination drew near. How often had she seen the tents of the English magistrate pitched there before, without a thought of him or of his business? . . . She seemed to have a direct concern with him. Would he be kind and considerate with her, or would he be rough and haughty, as she heard many English were?
>
> But she thought kindly of him. Every one spoke and thought well of him, she knew, and many were warm in his praises. He was courteous and just, she had often heard, and harmed no one.[40]

As the passage depicts Seeta's concern about her impending testimony, it also notes the space of the tent – which *stands in* for the colonial state and the rule of law – could to be a considerate and kind space or one marked by roughness and haughtiness. While these qualities are considered in the passage, the negative attributes are rejected with Seeta reassuring herself that Brandon was "courteous and just" and "harmed no one." In this process, the possibility of the colonial state space (signified by the "white tent") being rough, haughty or unjust is engaged with and rejected; instead it is now invested with notions of justice, courtesy, and kindness. Another structure of utopian colonial spatial desire can be seen in the depiction of Noorpoor during the first of January 1857.

> It was a pleasant day at Noorpoor; carriages and buggies were being driven about the station all day by callers; young men went "peacocking," as they called it, in their best and neatest trim; officials like the Brigadier, Mr. Mostyn, and Cyril Brandon, received complimentary visits from their subordinate officers and clerks; and trays of almonds and sugar-candy, with garlands of flowers, were presented by the two Serishtadars, the bankers of the bazaar,

> and the traders of the cantonment; and neat little speeches were made by the donors and recipients.[41]

This passage presents an image of happy conviviality with people dressing up and sharing gifts and engaging in active social intercourse. However, this scene of joyful co-mingling is necessarily undercut by a sense of dread and foreboding as the year, 1857, is clearly mentioned. The spatial utopianism is further continued in the depiction of the New Year's dinner party, which articulates a sense of home away from home for the British in India very clearly.

> The dishes were well served; a Portuguese butler carved the ham, and the courses went on regularly. Then the wines were almost iced by the mess "abdar," or water cooler; there was plenty of champagne, and the guests were happy.[42]

The idealization of these spaces is simultaneously marked with an intense fragility. One cannot *but* read these spaces as open to the possibility of being transformed where spaces invested with qualities of goodness could be taken over and transformed into spaces of evil.

The novel, in fact, delineates an alteration of this kind when it shows Azrael Pande and his hordes attacking the house of Huree Das and murdering him. And later in the novel when Pande's hordes attack the house where Seeta lives to abduct her, their attempts to transform that space are only thwarted by Brandon's bravery. Indeed, the novel sets up Brandon as a bulwark against the principle of evil and the attendant spatial desire:

> Cyril was standing in front near the steps, and . . . the Dacoits did not at once see him, but he had been concealed by one of the thick wooden pillars. Cyril let them rush in confusedly, and up the steps. He recognized the tall figure of the leader, though his face was muffled, and covering him with his pistol, fired; a man dropped, but it was not Azrael. Again and again Brandon fired rapidly, and in the confusion, which had become general – for the concealed men had rushed from their hiding-places, leaped into the court, and attacked the Dacoits.[43]

Brandon's heroism allows the novel to dramatize the trope of the "white saviour" and especially what Gayatri Spivak in her totemic formulation calls "white men . . . saving brown women from brown men."[44] But the anxiety of spatial transformation is of course most forcefully registered in the depiction of the actual troubles associated with the 1857 Uprising. In a depiction of the fighting, we are told:

> nearer and nearer came the bells, and suddenly some torchbearers appeared in the road behind Cyril's house, followed by a motley crowd of native foot soldiers, carrying bright matchlocks, and running at a steady pace. Then they

> dashed round the corner of the house, followed by an elephant, in the howdah of which some one was sitting . . .
>
> The sergeant, however, had got his aim perfectly. "Fire!" he cried, and the men ran to the parapet of the bastion to see the effect. The bursting of the shell hid the object for a moment; but an instant after, they saw the noble beast totter and fall.[45]

In this instance, unsurprisingly, the mutineers are thwarted and the colonial identity of Indian space, while substantially challenged, is nevertheless restored. It is significant that Taylor describes the killing of an elephant in this scene. The elephant gestures towards a structure of life that is very specifically Indian. And at the same time, it also provides a sense of the exotic and sensational – an elephant hunt – to its British readers.

Analogous to the anxiety that the quality of a space may change is the fear of a change in the nature of the subject. This is succinctly articulated in the novel's anxiety about Seeta. The novel repeatedly portrays Seeta as the object of Pande's sexual desire though, of course, she marries Brandon. Pande vows revenge on Seeta after his capture and as the narrative progresses, his desire is increasingly portrayed as sexualized until it becomes, after he is injured and ill (and thus qualitatively changed), into a kind of deranged and perverse sexual fantasy. He is shown saying:

> Seeta! Seeta! O beloved, come to me! . . . give me thy love, as thou hast mine! O lotos feet; I hear the sound of thy softly tinkling anklets! O lithe and swaying form advancing with dainty steps, I would embrace thee! O sweet Chumpa blossom, thy luscious perfume reaches me! I live, I drink it in! Seeta, I die! Come, touch me and this agony will cease.[46]

Azrael Pande's "impure" desire for Seeta is of course contrasted to the "pure" love of Brandon for her. He in fact does not keep her as his mistress as was the norm for colonial officers but ends up marrying her. This engenders a very different order of anxiety, especially among the British, which is articulated as unadulterated racism. This is evident in the way Mrs. Smith speaks about Seeta: "If men will have black 'companions,' you know, they ought to keep them to themselves, and not stuff them under our noses. I hate the thought of it. I feel quite sick as I look at them."[47] A few pages later, Mrs. Smith now speaking of Seeta's looks, notes that "she is as fair as any of us, and very pretty too . . . but she's a native, and they are all niggers, and – and – I hate them that's all."[48] This anxiety is paralleled, on the other hand, with Seeta descending "into a state of Gothic feared that Cyril means to abandon her, calling for protestations of loyalty and commitment from him."[49]

The narrative's anxiety around Seeta's status as the wife of Brandon, that is, the inter-racial marriage plot, is eventually resolved through her death at the hands of Azrael Pande. Pande too is killed in this melee thus effecting a symbolic catharsis of evil from the world of the narrative. This resolution is significant for several reasons. The narrative kills off Seeta and Azrael Pande – the twin sources of anxiety for the narrative – and, in so doing, ejects them from the world of the narrative. Seeta and Azrael Pande, in fact, cancel out each other thus gesturing towards an imaginary resolution of the real crises, namely, the Indian wife of a colonial British official (Seeta) and the mutinous and anti-colonial subject (Azrael Pande) in the symbolic realm. Seeta, furthermore, dies to save Brandon. And this attempt of the brown woman to save a white man from a brown man can be seen as a failed effort at reversing the "white saviour" trope. But while the narrative does resolve the twin crises of inter-racial marriage and anti-colonialism, this resolution is significantly different from the one outlined by Jameson in his discussion on the romance genre. For Jameson, the form of the romance effects a kind of "semic evaporation" where the category of evil represented through, for instance, a figure such as the hostile knight asks for mercy when defeated and is eventually "reinserted into the unity of the social class."[50] His analysis of the romance, however, does not engage with the peripheral context of the colonies where the racial divide underwrites the rule of capital and class antagonisms.[51] As we have seen, Seeta and Azrael Pande are not reinserted into the social structure; they are killed off and *removed* from the narrative world. Their deaths effect a resolution but it is not so much a *semic evaporation* as it is an *imaginary ejection* from the symbolic order and indeed the real world. This gestures towards a more violent attempt at pacifying the unruly natives that re-inscribes the fantasy of *terra nullius* on to the narrative world but now as the horizon of colonial utopianism. Such a response was hardly excessive for the colonizers, especially in the aftermath of the 1857 Uprising, and is succinctly encapsulated in Charles Dickens's letter from 4 October 1857 to Angela Burdett Coutts. Dickens writes:

> I wish I were Commander in Chief in India. The first thing I would do to strike that Oriental race with amazement . . . I should do my utmost to exterminate the Race upon whom the stain of the late cruelties rested; and that I was . . . now proceeding, with . . . merciful swiftness of execution, to blot it out of mankind and raze it off the face of the Earth.[52]

And Dickens's sentiment of colonial utopianism finds its most iconic, if laconic, expression in the literary field in Kurtz's fever-crazed imperative: "Exterminate all the brutes!"[53]

The anxiety of anti-colonialism: Bankim Chandra Chatterji's *Anandamath*

Anandamath appeared in serial form in the Bengali literary journal *Bongodorshon* [বঙ্গদর্শন], edited by the author himself in 1881, and was published as a book the following year. Immensely popular on publication, it was reissued in several subsequent editions throughout the remaining years of the nineteenth century.[54] The popularity of the novel, and especially the patriotic lyric *Vandemataram* [বন্দেমাতরম] within its pages, ensured the translation of the novel into English by the first decade of the twentieth century.[55] At the time of *Anandamath*'s publication,

Chatterji was already a pre-eminent literary figure in Bengal who was, owing to the centrality of the romance to his oeuvre, often compared to Walter Scott by the literary circle of nineteenth-century Calcutta. He was the editor of the much-admired *Bongodorshon*; his works had already been translated into English and were being published from Calcutta and London.[56] By 1910, Chatterji had entered the 11th edition of the *Encyclopedia Britannica* which described him as "beyond question the greatest novelist of India during the 19th century, whether judged by the amount and quality of his writings, or by the influence which they have continued to exercise."[57] The entry further notes that *Anandamath* was by far the most important of his novels, owing to its "astonishing political consequences."[58]

Anandamath is set in eighteenth-century rural Bengal, in the backdrop of the infamous famine of 1769–1773 that claimed the lives of 10 million people, which was about 30 per cent of the population of the region at that time.[59] The Bengal famine, called *chhiyattorer monwontor* [ছিয়াত্তরের মন্বন্তর] in Bengali, meaning "the famine of '76" after the year 1176 of the Bengali calendar, was the first of about 45 famines that South Asia witnessed during the 190 years of colonial rule between 1757 and 1947.[60] While the famine-afflicted landscape of Bengal forms the novel's backdrop, the text narrates the actions of a band of local insurgents called the *santan* [সন্তান] (literally, "the children") who have taken an oath of celibacy and organized themselves into a militant quasi-religious organization under their leader Satyananda. Despite their religious overtones, the *santan* have a political goal. They seek to overthrow the cruel and unjust regime of British revenue collectors of the English East India Company who in turn prop up a political order of the Mughal governor or *nawab* [নবাব] of Bengal. The nawab is depicted as weak and corrupt and the novel makes his rule *stand in* for Muslim rule of Bengal. The novel's initial equation of the English and the Muslim indicates these are versions of "bad" sovereignty but this also alludes to the history of eighteenth-century Bengal that was caught between the waning power of the Mughals and the British ascendancy in South Asia.

After the death of the Aurangzeb in 1707, the authority of the Mughal emperor in South Asia began to decline. This set up the conditions for the emergence of a number of regional rulers, all of whom ruled their regions in the name of the Mughal Emperor. The English and French East India Companies emerged as political players at this time. Through the mid-eighteenth to the early-nineteenth century, the English East India Company sought greater trading rights, especially rights to free trade, and in the process entered into alliances with some Indian rulers and waged wars against others.[61] Defeating in battle, over time, the nawab of Bengal, the French East India Company, Tipu Sultan of Mysore, the Marathas, and finally the Sikhs, the British established themselves as the premier power in South Asia with Calcutta, Bombay, and Madras as their seats. The British assumed power in Bengal in 1757, when they defeated Siraj-ud-daulah, the last independent nawab of Bengal, in the Battle of Plassey and installed Mir Jafar as a puppet *nawab*. In 1764, the English East India Company defeated the combined forces of the Mughal emperor, the nawab of Awadh, and the nawab of Bengal at the Battle of Buxar, winning the right to collect revenue from Bengal, then the richest province of the Mughal Empire. From 1765, Bengal was administered by a system of dual government with the British controlling revenue and taxation, and the nawab, the dispensation of justice. This system continued until the 1772 formation of the Bengal presidency, and the promotion of the Governor-General of Bengal to the Governor-General of India. The dual system of rule was abolished in 1793, when the British assumed complete control of Bengal. Company rule, that is, indirect colonial rule, of South Asia continued until the 1857 Uprising in the aftermath of which, Britain began to administer the British Indian colony directly.

I narrate this historical trajectory because *Anandamath* is set in the years around 1770, and refers to, and assumes the readers' familiarity with, Bengali political history of that time. The novel repeatedly underlines the devastation of Bengal owing to the English East India Company's revenue policies, and the *nawab's* abdication of his responsibility towards the welfare of the people. This allows it to equate the English East India Company and the Muslim rulers of Bengal, turning both into oppressors in a single move. Moreover, the *santan* state their political ambitions explicitly: to end Muslim and Company rule in Bengal and establish Hindu rule in its stead. In the course of the novel, the *santan* win numerous victories against the British and raise the cry of "vandemataram" (literally meaning, "I revere the mother") to celebrate each triumph depicting the space of Bengal, for the first time in public memory, as a "mother, ravaged by occupiers."[62] As the novel gained popularity, readers from outside the province of Bengal would take the spatial reference to signify India. Entwined within this larger narrative of *Anandamath* is the story of Mahendra, a landlord of the village Padacinha [পদচিহ্ন] and his wife Kalyani. Mahendra's wife and daughter are abducted by roving vagabonds in search of food, and eventually rescued by the *santan*. Mahendra is adopted into the *santan* fold, learns of their political goal, and helps make Padacinha into a fort from which they launch their final attack on the British forces. Another crucial figure in the narrative is Shanti, the wife of the celibate warrior-monk (Satyananda's follower) Jibananda who enters the "sacred brotherhood" at the abbey dressed as a man, the young monk Nabinananda. Kalyani and Shanti appear as foils in the narrative of *Anandamath*. If Kalyani is the meek and dutiful wife to her husband, Shanti is depicted as someone with a traditional education who rides horses and goes to battle against the British, recalling for the readers the warrior-queen of Monipur, Chitrangada, from the Sanskrit epic *Mahabharat*.

The novel opens with two striking instances of imagined geographies: one of the forest, and the other, of the village Padacinha ravaged by the Bengal famine. Let us examine these closely. At the very outset, readers are presented with the image of a forest in a prologue:

> A vast forest . . . The trees, with foliage intertwined, stretch out in endless ranks. Without break or gaps, without even openings for light to penetrate, a boundless ocean of leaves, wave upon wave ruffled by the wind, rolls on for mile after mile.
>
> Below, profound darkness prevails. Even at high noon the light is dim, dreadful! Humans never venture into that forest, and except for the ceaseless

> murmur of the leaves and the cries of its wild beasts and birds, no other sound is heard in it.[63]

In this extract, the text presents readers with the image of pristine nature untouched by humans. The forest is a space of lush vegetation and pastoral plenitude, filled with birds and beasts. It makes clear that humans "never venture into that forest," and no sounds are heard except for the rustle of leaves, and the cries of its non-human residents. The absence of human mediation suggests this space of the forest as one that exists beyond history. But the narrative goes back on this a few sentences later, telling us that the forest is not only not unmediated but profoundly marked by human presence: "In that boundless forest then, in that impenetrably dark night, in the inconceivable silence, a voice was heard: 'Will my heart's desire never be fulfilled?' "[64] And soon, the narrative underlines human presence in the forest further, stating that the space was the site of an "abbey" that gives the novel its title:

> There was a large monastery in that forest, spread out over a vast area and encircled by a great deal of broken stone . . . what had once been a Buddhist abbey had now become a Hindu monastery. There were two-storeyed rows of buildings with many kinds of shrines in between and a meeting-hall in front. Nearly the whole area was enclosed by a wall and covered in such a way by the thickets of trees outside that no one could tell even during the day and from a short distance that there were buildings here. . . . It was clear that even in this deep, *impenetrable* [দুর্ভেদ্য; durbheddyo] forest human beings now lived here.[65]

The manner in which history is being invoked in this passage is significant. Readers are told that what was once a "Buddhist abbey" has now become a "Hindu monastery." In commonsensical understandings of ancient South Asian history, Hindu kingdoms and rule are seen to have given way to a period of Buddhist rule. The site of the monastery not only draws on this historical narrative but reverses it. It is as if the space of the abbey bespeaks a desire to reclaim a lost past. Note, moreover, that the forest is presented as the staging ground for this speaker's "heart's desire." As the narrative unfolds, readers understand that the speaker is Satyananda, the leader of the *santan*, and his "heart's desire" is the liberation of the homeland from the British and Muslim rulers. It is therefore not surprising that the forest becomes a site from where their anti-colonial and anti-Muslim struggles are waged. Satyananda's "heart's desire" seems to also exude a spatial desire to transform the space of Bengal into a postcolony from a colony and set up a Hindu theocracy of sorts in its place.

But there are some more crucial aspects to the forest that need to be noted. First, the space of the forest *appears* natural and unmediated by human presence but, in *reality*, is actually profoundly shaped by it. The forest, then, is a site that presents a disjuncture between appearance and reality and is marked by a fundamental contradiction regarding the absence and presence of human mediation. This contradiction that structures the description of the forest permeates to the level of the sentence. We are told that "even in this deep, impenetrable [দুর্ভেদ্য; durbheddyo] forest human beings now lived here." Notice the use of the words "impenetrable" alongside the comment on human settlement. Surely if humans live in the forest, it cannot possibly be *impenetrable*. If we take the narrative's claims at face value, we would have to think that this space is both mediated by human presence and not; both part of human history and not – all this to say, simultaneously real and not, that is, an *impossible* space. Readers, unsurprisingly, are never told of the exact location of the forest, which retains a mythic and enchanted character throughout the novel. The forest signals the space of romance and is marked by what Mikhail Bakhtin calls "adventure time."[66]

The space of the famine-afflicted village of Padacinha, on the other hand, is depicted in stark contrast to the forest:

> It is summer one day in 1770. . . and the sun beats down fiercely. The village is full of homes, but there is no one about. There are rows of shops in the bazaar and lines of makeshift stalls in the marketplace, there are hundreds of mud houses in every quarter, with brick buildings of varying sizes in between, yet today everything is silent. The shops are closed, and no one knows where the shopkeepers have fled. It's the day for the local market, but the place is empty. The beggars are supposed to come out today, but none are about.
>
> The weaver has shut his loom and lies weeping in a corner, the trader has forgotten his trading and sobs with infant in his lap, the givers have stopped their giving, the teachers have closed their *tols* [schools], and even babies, it seems, lack the will to cry. No folk on the main roads, no bathers in the large ponds, no people at their doors, no birds in the trees, no cattle in the pastures – only jackals and dogs in the cremation ground.[67]

Unlike the forest, the depiction of the village appears to be historical, and realistic, in an overt and straightforward manner. Readers are presented with the social life of a quintessential Bengali village. The passage evokes the rural landscape through the "mud houses," "large ponds," and "*tols*" and signifies social life through another set of spatial markers, namely, "shops," the "bazaar," "stalls," and "marketplace" as well as figures such as "shopkeepers," "beggars," "weavers," and infants. The depiction of the space and social life here, as we have seen in the case of *Seeta*, also partakes the "affective impulse" of realist mode. But the space is also, however, presented as unnatural: not only as *it is* but also as *it should not be*. This sense is achieved through a series of negations: the village is full of homes but no one is about; there are signs of habitation but there is no sound of life; there are shops but there are no shopkeepers. This leads to the very powerful image of reversal: the village trader does not trade but sobs with an infant in his lap, and babies have lost the will to cry. Through this process of presenting a negational view of socio-spatial relations in Bengal, *Anandamath* presents the village of Padacinha as a world turned upside down.

It is noteworthy that *Anandamath* opens with the spaces of the forest and the village for it seems to signal them as complementary. The village space is marked by a realist impulse that describes it as a space of distress and of work – or, in these unnatural times heralded by the Famine of 1770, their absence. The novel blames the revenue demands of the British and the misrule of the nawab of Bengal for the famine. In so doing, it relates the distress of the village, and by extension, the larger space of Bengal, to the operations of the state. The space of the village for the narrative becomes a *stand in* for the state-space of the colony that is an outcome of, and an input to, the political-economic processes of colonial capitalism that produce and reproduce that space.[68] In addition, the space of the village also serves as a marker of time: it becomes a spatialization of the colonial present. This distinction between the *realist* space of the village and the *fabulous* space of the forest could be productively excavated through Jameson's observation that in the romance,

> in contrast to realism, its inner-worldly objects such as landscape or village, forest or mansion – mere temporary stopping places on the lumbering coach or express-train itinerary of realist representation – are somehow transformed into folds in space, into discontinuous pockets of homogenous time and of heightened symbolic closure, such that they become tangible analoga or perceptual vehicles for *world* in its larger phenomenological sense.[69]

Building on Jameson's observations, and translating it to the colonial context, we could say that the space of the forest in *Anandamath* signals a counter-colonial spatiality, or a space of counter-colonial possibility. This is the "good" space engendered by a counter-colonial spatial desire that stands in a relationship of determinate negation to the "bad" space signified by the village and produced by the operations of colonial capitalist modernity. I have noted earlier that the forest is presented as an impossible space. Now, I am suggesting that it gestures towards a possible counter-colonial space. The tension between the *impossible* and the *possible* animates the spatial character of the forest, and it is this simultaneity of the im/possible that marks another fundamental contradiction of the forest. It indicates the yoking together of a *not-here* in the *here* and the *not-yet* in the *now* resulting in the crystallization of an image of the *future non-place* in the *present space*. As such, the fabulous space of the forest signals the concretization of a counter-colonial desire into the emergent utopian space of the postcolony that is inchoate and can only be glimpsed as a structure of feeling.

As the narrative of *Anandamath* progresses, readers are told that the year 1770 came to an end, and the rains returned to Bengal the next year. But while the rains produced crops, people who were emaciated from lack of food were "unable to cope with a full stomach, and many died from this very condition."[70] This resulted in villages transforming into jungles:

> In village after village hundreds of fertile plots lay untilled and unproductive, or were covered with jungle. The whole land was filled with jungle. Where

> once rows of smiling dark green crops had graced the land and countless cows and buffaloes had grazed, where parks had once been the dallying-grounds of village youths and maidens, now dense jungle gradually began to grow.
>
> Three whole years passed, and the jungle proliferated. Places where humans had lived happily now saw man-eating tigers pursue deer and other prey. Where once groups of beautiful women, their anklets tinkling on lac-painted feet, had made their way joking and laughing aloud, now bears made their lairs and reared their young. Where once little children had blossomed like the evening jasmine and laughed to their hearts' content, now herds of rutting wild elephants tore the tree trunks. The place of the Goddess Durga's festival now became the jackal's lair, the dais for Krishna's Dol festival became the owl's refuge, while in the temple's meeting hall poisonous snakes searched for frogs in broad daylight.[71]

This extended extract from *Anandamath* points to a transformation in the spatial character of the village. Readers are told that owing to the death of its residents, the villages were taken over by the jungle. This is crucial, as it not only illuminates the continuing dichotomy between the village space and the space of the forest, but also shows the former giving way to the latter. The village is once again presented as a world turned upside down and described in an oppositional relationship to the forest. Notice, for instance, how the passage constructs oppositions between "humans" and "man-eating tigers"; between "beautiful women" and "bears"; between "little children" and "rutting wild elephants." This is paralleled through another set of oppositions between human actions and the actions of wild animals: the humans living happily vis-à-vis tigers pursuing "deer and other prey"; women making "their way, joking and laughing aloud" vis-à-vis bears rearing "their young"; the laughter of children vis-à-vis elephants tearing tree trunks. The proliferation of the jungle also results in the overall transformation of the space as a site of dwelling. The site of the annual festival of the Durga turns into "the jackal's lair," the platform for *doljatra* [দোলযাত্রা; "dol festival"], an owl's nest, and the temple's hall, a lair of snakes. The passage does not simply record a transformation but also a change in the spatial quality of the village. The space of the village, which we have earlier encountered as a historical space of affective realism, has now taken on a fabulous quality identified earlier with the forest. It shows the colonization of the historical world of the village by the forest, which signals a mythical world. To put this another way, the disenchanted world of colonial capitalist modernity, marked by the ravages of famine, has been re-enchanted and taken over, at the level of narrative, by the space of romance. In so doing, the narrative exudes a spatial desire that takes on the structure of wish-fulfilment and presents the village as an altered space.

The utopianism of the spatial imaginary of *Anandamath* is further underlined in the text's inauguration of a radically novel understanding of space in a temporal vein. This is carried out through a gendering and deification of the space of Bengal, and a concomitant spatialization of gender and divinity, through the deployment of the images of the mother goddess.[72] Much of the labour of this ideological move of gendering and deifying space is carried out by the patriotic lyric *Vandemataram* (I discuss it in chapter 3) but the novel's articulation of this deified and gendered space as *temporalized* is just as significant. Most crucially, it gestures towards the inseparability of time and space. These depictions allow time to, as it were, thicken and take on flesh while illuminating space as "responsive to the movements of . . . history."[73] I draw on Bakhtin's formulation of the chronotope here for it illuminates these images of temporalized space as chronotopes constructed using the "common-sense" of popular Hinduism and its symbols. But *Anandamath* does more than simply fashion Hindu chronotopes; it uses them to present space and time through the categories of "good" and "bad." It deploys them to articulate the notion of progress and inscribe a sense of *telos* into the history of space. In other words, it devises a progressive history for Bengal (and, by extension, India).

In a central moment in the narrative, Satyananda takes Mahendra into the temple of the *santan* in their abbey. Mahendra is shown an image of the Hindu god Vishnu, and on "Vishnu's lap sat an enchanting image."[74] This image, Mahendra is told by the monk, is "the Mother" whose children they all were. Mahendra is subsequently taken to another room and shown the image of Jagaddhatri [জগদ্ধাত্রী; literally, "the bearer of the earth"] and is told that this is the image of the *Mother-as-she-was* [মা যা ছিলেন].[75] Mahendra is then taken to a "dark chamber, in the depths of the earth, lit somehow by a faint light" where he encounters the image of Kali [কালী].[76] This, Satyananda tells Mahendra, is the *Mother-as-she-is* [মা যা হইয়াছেন].[77] Finally, Mahendra and Satyananda ascend through a tunnel to a sanctum where "the morning sun dazzled their eyes and they heard the soft bird-song all around them. Mahendra saw a golden ten-armed image of . . . [Durga; দুর্গা] in a large marble shrine glistening and smiling in the early morning rays."[78] And this image signifies, Satyananda tells Mahendra, the future *Mother-as-she-will-be* [মা যা হইবেন]. Note how these images of Jagaddhatri, Kali, and Durga collapse the categories of gender, divinity, space, and time. In this formulation, Jagaddhatri signifies a gendered and deified image of past space; Kali the space of the present, and Durga, the space of the future. The reliance on these icons of the mother goddess to depict the space presents Bengal in unabashedly Hindu terms. This would make some later translators uneasy with the novel's vision of Bengal (and India) as a Hindu space. For instance, the 1941 translation of *Anandamath* by Basanta Koomar Roy titled *Dawn over India* substitutes the idols of the mother goddesses with maps of India. The space of the past becomes "a gigantic, imposing, resplendent . . . almost a living map of India"; the present is signified through a map of India "in rags and tears," and the space of future through a map of "golden India – bright, beautiful, full of glory and dignity."[79] And note also how in signalling the space as India, the translation points to the naturalization, by the 1940s, of reading the spatial imaginary of *Anandamath* as an allegory of the nation.

The use of the specific icons of the mother goddess is, nevertheless, significant as it points to an aestheticization of spatial politics.[80] For instance, Jagaddhatri, the deity understood to be the bearer of the earth, is, to the popular imagination, the "beautiful" deity who creates and nourishes the universe. The deployment of Kali is even more noteworthy as it is the "sublime" form of the mother goddess.[81] She is traditionally imagined as terrible, and one who inspires fear and awe. Conceived of as wandering the cremation grounds, she holds a severed head in one hand, and a scimitar in another. She is also formally depicted as bare-chested, wearing a garland of human heads, and a frock of human hands across her waist (see Figure 1.1). Most significantly, she is always shown stepping on the male god, and her husband, Shiva. The word "kali" [কালী] derives from

Figure 1.1 Popular depiction of the Hindu goddess Kali

Source: From the personal collection of Subhashish Roy.

"time" [কাল; *kal*], while the word "shiva" [শিব; *shib*] literally means "well-being." The goddess' act of stepping on her husband is popularly interpreted as an inversion of the cosmic order that makes it a non-western version of the "woman on top" trope that challenges, and indeed overturns, the patriarchal social order.[82] The use of the image of Kali, then, allows the narrative to achieve two ends. Foremost, it conveys a sense of the devastation of the land under colonization in an affective vein. The Kali image also speaks of, and to, a male anxiety regarding the space of Bengal and the attendant socio-spatial relations. It is therefore unsurprising that the mother's male *santan* strive to transform the space, and the mother-image associated with it, to the more "picturesque" Durga.[83] Durga is the most popular Hindu goddess of Bengal, and is traditionally imagined with ten hands, bearing ten weapons of war, whose most notable exploit according to Hindu scriptures was the slaying of the male warrior-king Mohishashur. Though she is formally venerated as a warrior-goddess and a slayer of evil, the social affect surrounding her annual festival in Bengal is of a more domestic, and domesticating, tenor.[84] In Bengal, she is popularly understood as a married daughter from the region, and the annual festival in her honour is understood as the time she visits her natal home (Bengal) from her husband's (the Himalaya). In addition, she is also shown visiting Bengal with her two daughters and two sons and is worshipped *en famille* with her children (see Figure 1.2). While Durga symbolizes feminized political agency, that image of hers is tamed by the more immediate and affective one of familial domesticity that presents her as the ideal daughter and mother. This makes her, unlike Kali, an unthreatening female divinity. Naturalizing her as the embodiment of the future space also corrects the cosmic (and social) inversion, reaffirming the "natural" patriarchal order, reversing the "woman on top" trope.

This understanding of progressive history points, necessarily, towards decolonization as its utopian horizon. *Anandamath* makes it clear that the *santan* understand decolonization as the end of British and Muslim rule, and the establishment of Hindu rule in Bengal. This utopian vision of the narrative, however, pushes its resolution towards crisis. By the end of the novel, the *santan* win a major battle against the British. However, despite providing this imaginative gloss to the real conditions of existence, the narrative cannot alter the events of the world beyond the symbolic action of the text. It can show British defeat *within* the narrative but cannot end British rule, or establish Hindu rule, in South Asia. It resolves the crisis between promise and actuality though a *deus ex machina*. It devises a mysterious seer – Satyananda's guru – who urges military restraint upon him and his followers. The seer tells Satyananda, "Your work's been done, and Muslim rule has been destroyed. There's no more for you to do."[85] And when Satyananda replies that "Hindu rule has not been established. Even now the English remain powerful in Kolkata," the seer prophesizes British rule as a necessary period in the history of India.[86] He says:

> the outward knowledge has been lost in this land, and so the true Eternal Code [সনাতন ধর্ম; the Hindu way of life] has been lost too. If one wishes to reinstate

Figure 1.2 Durga being worshipped *en famille* in South Calcutta

Source: Image courtesy Arijit Sen.

this Code, one must make known the outward knowledge first. The outward knowledge no longer exists in this land, and there's no one to teach it; we ourselves are not good at teaching people such things. So we must bring in the outward knowledge from another country. The English are very knowledgeable in the outward knowledge, and they're very good at instructing

> people. Therefore we'll make them king. And when by this teaching our people are well instructed about external things, they'll be ready to understand the inner . . . so long as the Hindu is not wise and virtuous and strong once more – English rule will remain intact.[87]

This resolution achieves several ends for the narrative. First, it sets up what Chakrabarty has, in his discussion on John Stuart Mill's work on self-government of the colonies, called the colonizer placing the colonized in the "waiting room" of history.[88] This extract from *Anandamath* shows the internalization and justification of that principle. In addition, it drops its anti-colonial outlook entirely, and presents a historical development – the establishment of British rule in South Asia – as a function of historical necessity. In so doing, *Anandamath* draws upon colonial historiography of India pioneered by James Mill, James Todd, Henry Elliot, and Mountstuart Elphinstone.[89] These positivist accounts of Indian history presented British rule as the teleological end in the process of political domination of the Indian subcontinent, progressively by Hindus, Muslims, and the British. Moreover, these histories presented Muslims as tyrannical foreign conquerors of India. In addition, the historical vision of *Anandamath* is, however, not simply structured by this colonial historiographical tradition. By projecting a future spatio-temporal reality of India under Hindu rule, it also uses the colonial logic of progressive history against itself. In addition, a structure of wish-fulfilment is also evident in this extract. Note the mysterious seer saying, "we'll make them [the British] king." This is instructive for it shows a reversal of causality, and a wishful ascription of agency to the Bengalis (and Hindus), in the establishment of British power in South Asia. The most significant aspect of *Anandamath*'s conclusion, however, is its letting go of the counter-colonial desire. Instead of seeking to get rid of the British from Bengal and establish Hindu rule there, the narrative directs its entire utopian charge at the ending of Muslim rule in Bengal. Thinking about this through the categories of "good" and "bad" of the romance form, we could say that at the start of the narrative, *Anandamath* presents a future Hindu space as a "good" version of the space of Bengal, while that space under British and Muslim rule are seen as versions of the "bad." As the narrative progresses, British and Muslim rule are decoupled. By the time of the narrative's denouement, British rule in Bengal is magically transformed from "bad" to the "good," with only Muslim rule in Bengal being presented as a version of "bad." This necessarily changes the nature of the counter-colonial spatial desire from anti-British and anti-Muslim to *only* anti-Muslim. In other words, the narrative moves from an anti-British to a pro-British position by its end. Sudipta Kaviraj suggests that this conclusion to the narrative makes it function in a manner that is

> exactly the opposite of what Lukacs found in the European historical novel. . . . The point of these novels is precisely to 'falsify' history. They try deliberately to probe and use counterfactuals, to extend those lines in the tree of eventuation which were not actually followed up, explore the peculiar terrain of history's nonactualized possibilities.[90]

This is a somewhat curious and tendentious reading since the closure, even while acknowledging the non-actualization of Hindu rule, does present – and justify – the *actualization* of British rule, which Kaviraj's reading simply wishes away. Instead of falsifying history, the structure and closure of *Anandamath* allows it to take on a prophetic tone. Recall, the novel was published in 1881 in the journal *Bongodorshon*, and as a novel in 1882, but set in the years around the Great Famine of 1770. This encodes the narrative with a sense of historical hindsight: the *evolving present* of the narrative is the *historical past* of its readers.[91] This temporal disjuncture, along with the *deus ex machina* of the closure, allows the narrative of *Anandamath* to articulate a secular theodicy; that is to say, not so much as justifying the ways of God to men but explaining the ways of history to the colonized subjects, and readers.

The notions of "good" and "bad that I have been delineating in *Anandamath* with respect to its spatial imaginaries also extend to the construction of the identities of individuals and, in particular, social groups. The question of "good" and "bad" in the colonial context of the romance can also be seen as attempts to define – and settle – questions of belonging. This becomes particularly important for *Anandamath* for thinking of the status of Muslims in the space of Bengal. Chatterji creates the normative identity of the Bengali as a Hindu, turning the minority figure into a hegemonic one. This is evident in the novel's description of the aftermath of the battle where the identities of "Bengali" and "Muslim" are suggested to be distinct: "Bengali, Northern or, English, Muslims are strewn in each other's embrace; the living, the dead, man and horse, light jumbled together."[92] Indeed, the Muslim becomes a figure of evil in *Anandamath*. As I have noted, the spatial desire of the novel moves, by the time of its ending, from being anti-British and anti-Muslim to being focused solely on ending Muslim rule in Bengal. This runs parallel to the novel's anxiety about Muslims within, and its desire to evict them from, the space of Bengal. During the narrative, when the *santan* attack the jail to free their leader Satyananda, one of the warrior monks says:

> For a long time we've been wanting to smash the nest of these weaver-birds, to raze the city of these Muslim foreigners [যবন; *jabon*], and throw it into the river – to burn the *enclosure of these swine* and purify Mother Earth again! Brothers, that day has come![93]

Note here the equation of the Muslim residents to "the nest of these weaver-birds" and, more crucially, to them as pigs. Also note how the Muslim is being presented as a foreigner. As I mentioned, this shows Chatterji's internalization of colonial historiography that portrayed Muslims as outsiders to India, while demonizing Muslim rule as debased to justify their own colonial enterprise. The spatial desire of *Anandamath* seeks to eject the Muslim from the space of Bengal (and India), in effect, striving to construct what Anupama Mohan calls a "homotopia," that is, "visions of unified collectivity where an aggressively homogenizing impulse operates and where unity is a form of collective gathering of one or two coordinates (race/language/religion) and the deliberate repudiation and exclusion of others."[94] In another scene from the novel, the jubilation of the *santan* also takes on an anti-Muslim tenor:

> Some began to shout, "Kill, kill the baldies!" [মার, মার, নেড়ে মার] . . . Some of the men said to one another, "Brother, has the day come when even the insignificant Bengali can lay down his life on the battlefield?"; others said, "Brother, has the day come when we will be able to tear down the mosque and build a temple to Radha and Krishna in its place? [ভাই, এমন দিন কি হইবে, মসজিদ ভাঙ্গিয়া রাধামাধবের মন্দির গড়িব?]"[95]

Some terminological explanation is required here: "baldy" is the English translation of the Bengali word *nerey* [নেড়ে], a slur used to describe the Muslim, which refers to the shaved heads of some members of the religious community. And in "the insignificant Bengali laying down his life on the battlefield," we see the internalization of another colonial trope: the weak and effeminate Bengali who is sly, cunning, and intelligent but not gifted like the Pathan or the Punjabi in martial valour.[96] The narrative repeatedly equates the British colonizer with the Muslim, turning them both into versions of the "bad" and its delineation of subjects follows the trajectory of outlining the characteristics of space. By the end of the narrative, the British become a version of the "good" while the narrative's Muslim remains consigned to the category of "bad." *Anandamath*'s imagination of the postcolony stand in an oblique relationship of overlap and contestation with the spatial imaginary of *Seeta*. These two romances highlight two different iterations of spatial desire in late-nineteenth-century British India. *Anandamath*'s imagination of the Muslim subject, and the space of Bengal (and India) under Muslim rule, would have far-reaching consequences. It would popularize a certain view of the Muslim as "bad," and of Muslim rule in India as "debased" and "debauched" that would inform and animate the right-wing Hindu polity in colonial, as well as postcolonial, India.

Notes

1 Northrop Frye, *The Secular Scripture: A Study of the Structure of Romance* (Boston: Harvard University Press, 1976), 15. Also see Northrop Frye, *The Anatomy of Criticism: Four Essays* (Princeton, NJ: Princeton University Press, 1971), 186–206.

2 Fredric Jameson, *The Political Unconscious: Narrative as a Socially Symbolic Act* (Ithaca: Cornell University Press, 1982), 96–97.

3 Ibid., 97–106.

4 Ibid., 105.

5 The scholarship on the romance form in South Asian literature remains extremely limited. Notable exceptions include Shuchi Kapila, *Educating Seeta: The Anglo-Indian Family Romance and the Poetics of Indirect Rule* (Columbus: Ohio State University Press, 2010); Sangeeta Ray, *En-Gendering India: Woman and Nation in Colonial and Postcolonial Narratives* (Durham, NC: Duke University Press, 2000); and Henry Schwarz, "Sexing the Pundits: Gender, Romance, and Realism in the Cultural Politics of Colonial Bengal," in *Reading the Shape of the World: Toward an International Cultural Studies*, eds. Henry Schwarz and Richard Dienst (Boulder: Westview Press, 1996), 224–258.

6 "Anandamath" literally means "the abbey of joy or bliss." The English translations of *Anandamath* have rendered the title, variously, as *The Abbey of Bliss*, *Dawn over India*, and *Sacred Brotherhood.*

7 Henri Lefebvre, *The Production of Space*, trans. Donald Nicholson-Smith (Oxford: Blackwell, 1991), 26.

8 Ernst Bloch, Theodore Adorno, and Horst Kruger, "Something's Missing: A Discussion Between Ernst Bloch and Theodor W. Adorno on the Contradictions of Utopian Longing," in *The Utopian Function of Art and Literature: Selected Essays*, trans. Jack Zipes and Frank Mecklenburg (Cambridge: MIT Press, 1989), 16.

9 This cathartic and ritualistic aspect of the text has been taken up extensively in studies on melodrama. See, for instance, Peter Brooks, *The Melodramatic Imagination: Balzac, Henry James, Melodrama, and the Mode of Excess* (New Haven: Yale University Press, 1976); Linda Williams, "Melodrama Revised," in *Refiguring American Film Genres: History and Theory*, ed. Nick Browne (Berkeley: University of California Press, 1998), 42–88; and Christine Gledhill, "The Melodramatic Field: An Investigation," in *Home is Where the Heart Is*, ed. Christine Gledhill (London: British Film Institute, 1987), 5–39.

10 For a discussion on the symbolic resolution in narrative of real contradictions see, in addition to Jameson, *Political Unconscious*, the chapter "Reification and Utopia in Mass Culture" in Fredric Jameson, *Signatures of the Visible* (New York: Routledge, 1990), 11–46 and René Girard, *Deceit, Desire, and the Novel: Self and Other in Literary Structure* (Baltimore: Johns Hopkins Press, 1965).

11 For studies on the 1857 Uprising, see Kazi M. Ashraf and Puran C. Joshi, *Rebellion, 1857: A Symposium* (New Delhi: People's Publishing House, 1957); Eric Stokes and C. A. Bayly, *The Peasant Armed: The Indian Revolt of 1857* (Oxford: Clarendon Press, 1986); and Crispin Bates and Marina Carter, *Mutiny at the Margins: Documents of the Indian Uprising* (New Delhi: Sage Publications, 2017). For a study on the 1857 Uprising on the British Imagination, see Gautam Chakravarty, *The Indian Mutiny and the British Imagination* (Cambridge: Cambridge University Press, 2005).

12 Zahid Chaudhuri, "Phantasmagoric Aesthetics: Colonial Violence and the Management of Perception" *Cultural Critique* 59 (2005): 67.

13 John Murray was commissioned to photograph the important sites of the 1857 Rebellion by Charles Canning, the last British Governor General and first Viceroy of India. The gallery of J. Hogarth published these images of Murray in 1859 as *Picturesque Views in the North Western Province of India.* In 1859, Felice Beato's images of the Rebellion were published as woodcuts by *The Illustrated London News*. In 1862, London-based photographer, print dealer, and publisher H. Hering began selling catalogues comprising Beato's images. For details see Joachim K. Bautze, "The Beginning of Photography in India and the Photographers in *Picturesque Views*," in *Picturesque Views: Mughal India in Nineteenth-Century Photography*, ed. Raffael Dedo Gadebusch (Ostfildern: Hatje Cantz, 2007), 21–26.

14 See G. Harris, *A Lady's Diary of the Siege of Lucknow: Written for the Perusal of Friends at Home* (London: J. Murray, 1858); Robert D. Gibney, *My Escape from the Mutinies in Oudh* (London: Richard Bentley, 1858); and Mowbray Thompson, *Story of Cawnpore* (London: Richard Bentley, 1859).

15 See Charles Ball, *The History of the Indian Mutiny: Giving a Detailed Account of the Sepoy Insurrection in India: And a Concise History of the Great Military Events Which Have Tended to Consolidate British Empire in Hindostan* (London: London Print and Pub. Co., 1858); and George Trevelyan, *Cawnpore* (London: Macmillan and Co., 1866).

16 Patrick Brantlinger, *Rule of Darkness: British Literature and Imperialism, 1830–1914* (Ithaca: Cornell University Press, 1988), 200.

17 Philip Meadows Taylor, *Seeta* (London: C. Kegan Paul & Co., 1880), 69. My italics.

18 Ibid., 68.
19 Edward Said glosses the word "orientalism" in three ways in his classic study. The first, the sense in which I use it here, refers to someone who studies the orient. The second signals an epistemological and ontological distinction between the orient and the occident while the third sense, which forms the thrust of his book and its argument, is an extended exercise of positional superiority in political and cultural registers (he is largely silent about the economic domain) through which the occident defines and dominates the orient. See, Edward Said, *Orientalism* (New York: Vintage Books, 1979), 1–28.
20 Taylor, *Seeta*, 69.
21 Ibid., 67.
22 Ibid., 69.
23 Kapila, *Educating Seeta*, 3.
24 Taylor, *Seeta*, 61. My italics.
25 Kapila, *Educating Seeta*, 3.
26 Ibid., 117.
27 Taylor, *Seeta*, ix.
28 Ibid., 4–5. My italics.
29 Ibid., 1.
30 See Fredric Jameson, *The Antinomies of Realism* (London: Verso, 2013), 27–44.
31 Taylor, *Seeta*, 1–2.
32 Ibid., 2.
33 E. M. Forster, *A Passage to India* (Harmondsworth: Penguin Books, 1924), 11.
34 Taylor, *Seeta*, 2.
35 Ibid., 3.
36 Ibid.
37 Ibid., 16.
38 Ibid.
39 Ibid., 17.
40 Ibid., 58–59.
41 Ibid., 183.
42 Ibid., 17.
43 Ibid., 103.
44 Gayatri Chakravorty Spivak, "Can the Subaltern Speak?," in *Marxism and the Interpretation of Culture*, eds. Cary Nelson and Lawrence Grossberg (Urbana: University of Chicago Press, 1988), 296.
45 Taylor, *Seeta*, 351.
46 Ibid., 300.
47 Ibid., 182.
48 Ibid., 183.
49 Kapila, *Educating Seeta*, 125.
50 Jameson, *The Political Unconscious*, 105.
51 It must be noted that in his discussion of *Jaws* in the essay "Reification and Utopia in Mass Culture," Jameson reads the death of Quint as an allegory for "the twofold symbolic destruction of an older America . . . of small business and individual private enterprise of a now outmoded kind [and] . . . the America of the New Deal and the crusade against Nazism, the older America of the depression and the war and of the heyday of classical liberalism." See "Reification and Utopia," 38. This, however, does not take away from my larger point about the narrative resolution in the colonial romance.
52 Quoted in Brantlinger, *Rule of Darkness*, 206–207.
53 Joseph Conrad, *Heart of Darkness*, ed. Robert Hampson (London: Penguin Books, 2007), 62.

54 The second edition of *Anandamath* was published in 1883 with a print run of 1,000 copies. Its third and fourth editions were published in 1886 with print runs of 1,000 and 2,000 respectively. A fifth edition of 1,000 copies was issued in 1892. The fifth edition is typically taken as the standard edition of *Anandamath.* For details, see Julius Lipner, *Introduction to Anandamath or The Sacred Brotherhood,* by Bankim Chandra Chatterji, trans. Julius Lipner (New York: Oxford University Press, 2005), 34–36.

55 *Anandamath* was by first translated into English by Nares Chandra Sen Gupta; its fifth edition was published from Calcutta in 1906 as *The Abbey of Bliss.* In 1909, another translation by the Indian nationalist-turned-philosopher Aurobindo Ghose, and his brother Barindra Kumar Ghose, was published from Calcutta. The third English translation by Basanta Koomar Roy, titled *Dawn over India,* was published from New York in 1941. See, Lipner, introduction, 44.

56 For instance, see Bankim Chandra Chatterjee, *Durgesa Nandini or, The Chieftain's Daughter: A Bengali Romance*, trans. Charu Chandra Mookerjee (Calcutta: H. M. Mookerjee & Co., 1880); Bankim Chandra Chatterjee, *The Poison Tree: A Tale of Hindu Life in Bengal*, trans. Miriam S. Knight (London: T. Fischer Unwin, 1884); Bunkim [sic] Chandra Chatterjee, *Kopal-Kundala: A Tale of Bengal Life*, trans. H. A. D. Phillips (London: Trübner & Co., 1885); *and* Bankim Chandra Chatterjee, *Krishna Kanta's Will*, trans. Miriam S. Knight (London: T. Fischer Unwin, 1895).

57 *The Encyclopaedia Britannica*. 11th ed., vol. 6. "Chatterji, Bankim Chandra" (Chicago: Encyclopedia Britannica, 1910).

58 Ibid.

59 Leela Visaria and Pravin Visaria, "Population, 1757–1947," in *Cambridge Economic History of India, vol. 2: c.1757–1970*, eds. Dharma Kumar and Meghnad Desai (Cambridge: Cambridge University Press, 1983), 527.

60 B. M. Bhatia, *Famines in India; A Study in Some Aspects of the Economic History of India, 1860–1965* (Bombay: Asia Publishing House, 1967), 8. The most proximate famine to the novel's date of publication was the famine of 1876–1878, this time in southern India, killing about 5 million people. Ironically enough, the last major famine in colonial India was also in Bengal. Lasting from 1943–1944, modest estimates put the death toll at 3 million. For studies on famines in colonial India see Amartya Sen, *Poverty and Famines: An Essay on Entitlement and Deprivation* (Oxford: Oxford University Press, 1983) and Janam Mukherjee, *Hungry Bengal: War, Famine and the End of Empire* (New York: Oxford University Press, 2015). For a global history of famines during the Victorian era, see Mike Davis, *Late Victorian Holocausts: El Niño Famines and the Making of the Third World* (London: Verso, 2002). For the literary registration of natural calamities, including famines, in the Victorian era, see Upamanyu Pablo Mukherjee, *Natural Disasters and Victorian Empire: Famines, Fevers and the Literary Cultures of South Asia* (Basingstoke: Palgrave Macmillan, 2013).

61 See Sudipta Sen, *Empire of Free Trade: The East India Company and the Making of the Colonial Marketplace* (Philadelphia: University of Pennsylvania Press, 1998).

62 Priya Joshi, *In Another Country: Colonialism, Culture, and the English Novel in India* (New York: Columbia University Press, 2002), 161.

63 Bankim Chandra Chatterji, *Anandamath, or, the Sacred Brotherhood*, trans. Julius Lipner (New York: Oxford University Press, 2005), 129. For the Bengali text of the novel, see Bankim Chandra Chatterji, *Anandamath* [আনন্দমঠ], ed. Brajendranath Bandyopadhyay and Sajonikanto Sen (Calcutta: Bangiya Sahitya Parishat, 1938).

64 Chatterji, *Anandamath,* 129.

65 Ibid., 137.

66 See Mikhail Bakhtin's discussion on the "Greek Romance," in M. M. Bakhtin, "Forms of Time and the Chronotope in the Novel," in *The Dialogic Imagination: Four Essays by M. M. Bakhtin*, ed. Michael Holquist, trans. Caryl Emerson and Michael Holquist (Austin: University of Texas Press, 1981), 86–110.

67 Chatterji, *Anandamath*, 130.
68 See Manu Goswami, *Producing India: From Colonial Economy to National Space* (Chicago: University of Chicago Press, 2004), 31–72.
69 Jameson, *The Political Unconscious*, 98.
70 Chatterji, *Anandamath*, 188.
71 Ibid.
72 Sangeeta Ray provides insightful readings of female characters of the novel such as Shanti and Kalyani. She does not comment on the gendering of space in *Anandamath*. See Ray, *En-Gendering India*, 39–48. Also see Tanika Sarkar, *Hindu Wife, Hindu Nation: Community, Religion, and Cultural Nationalism* (Bloomington: Indiana University Press, 2001).
73 Bakhtin, "Forms of Time," 84.
74 Chatterji, *Anandamath*, 149.
75 Ibid.
76 Ibid., 150.
77 The Bengali phrase *ma ja hoiyachen* [মা যা হইয়াছেন] is translated as "Mother-as-she-is." While this translates, literally, into "mother-as-she-has-become," it also conveys the sense of "the state to which the mother has been reduced."
78 Chatterji, *Anandamath*, 150.
79 Bankim Chandra Chatterji, *Dawn over India*, trans. Basanta Koomar Roy (New York: The Devin-Adair Co., 1941), 62–63. This version also translates the *santan* leader Satyananda as "Mahatma Satya" in what appears to be an allusion to Mohandas Gandhi, popularly referred to as Mahatma, who at that time was the undisputed leader of the Indian anti-colonial struggle. Unlike Gandhi, however, Mahatma Satya of the novel notes that India "can be freed only by the sword. Those that talk of winning India's freedom by peaceful means do not know the British." See Chatterji, *Dawn over India*, 62–63.
80 For a distinction between the "aestheticizing of politics" and, conversely, the "politicizing of art" see Walter Benjamin, "The Work of Art in the Age of Mechanical Reproduction," in *Illuminations*, ed. Hannah Arendt, trans. Harry Zohn (New York: Schocken Books, 1969), 242.
81 I am here drawing on the important categories of the "beautiful," "sublime," and "picturesque" from western aesthetics. For an explication of the categories, see James Buzzard, "The Grand Tour and After (1660–1840," in *The Cambridge Companion to Travel Writing,* eds. Peter Hulme and Tim Youngs (Cambridge: Cambridge University Press, 2002), 37–52. For a delineation on the category of the sublime, see Immanuel Kant, *Critique of Judgment*, trans. J. C. Meredith (Oxford: Clarendon Press, 1952). Also see Edmund Burke, *A Philosophical Enquiry into the Origin of Our Ideas of the Sublime and Beautiful* (London: Routledge and Paul, 1958).
82 For a discussion of the "woman on top" trope, although in the context of early modern France, see Natalie Zemon Davis, *Society and Culture in Early Modern France: Eight Essays* (Stanford: Stanford University Press, 1975), 124–151.
83 For an elaboration on the idea of the picturesque, see William Gilpin, *Three Essays: On Picturesque Beauty; On Picturesque Travel; And on Sketching Landscape: To which is Added a poem, on Landscape Painting* (London: Printed for R. Blamire, 1792).
84 Tithi Bhattacharya suggests that in her normative Hindu conception, Durga is already domesticated and rendered brahminical. She notes that the Sanskrit texts that mention her before her incorporation into the pantheon, such as the *Mahabharat* and *Harivamsa,* indicate her non-brahminical provenance. She is described in those texts as "a fearless virgin" dressed in peacock feathers who hunts and lives in the mountains. She has ghosts and wild beasts for companions and consumes "meat and alcohol" considered taboo by brahminical Hinduism. See Tithi Bhattacharya, "Tracking the Goddess: Religion, Community, and Identity in the Durga Puja Ceremonies of Nineteenth-Century Calcutta," *The Journal of Asian Studies* 66 (2007): 925.

85 Chatterji, *Anandamath*, 228.
86 Ibid.
87 Ibid., 229.
88 Chakrabarty, *Provincializing Europe*, 8.
89 For colonial history writing, see Javed Majeed, *Ungoverned Imaginings: James Mill's The History of British India and Orientalism* (Oxford: Clarendon Press, 1992).
90 Sudipta Kaviraj, *The Unhappy Consciousness: Bankimchandra Chattopadhyay and the Formation of Nationalist Discourse in India* (New Delhi: Oxford University Press, 1995), 133.
91 This narrative device is not unlike Books XI and XII of John Milton's *Paradise Lost* or the vision of the future glory of Rome presented to Aeneas by the Cumaean Sibyl in Book VI of Virgil's *Aeneid*. See John Milton, *Paradise Lost*, ed. Alastair Fowler (Harlow: Longman, 2007), 597–698; and Virgil, *Aeneid*, trans. Frederick Ahl (Oxford: Oxford University Press, 2007), 128–157.
92 Chatterji, *Anandamath*, 226.
93 Ibid., 169. My italics.
94 Anupama Mohan, *Utopia and the Village in South Asian Literatures* (Basingstoke: Palgrave Macmillan, 2012), 8–9.
95 Chatterji, *Anandamath*, 203.
96 For a discussion on Chatterji's construction of Hindu masculinity, see Chandrima Chakraborty, *Masculinity, Asceticism, Hinduism: Present and Past Imaginings of India* (Ranikhet: Permanent Black, 2011), 39–78; also see, Sikata Banerjee, *Make Me a Man!: Masculinity, Hinduism, and Nationalism in India* (Albany: State University of New York Press, 2005); and Indira Chowdhury, *The Frail Hero and Virile History: Gender and the Politics of Culture in Colonial Bengal* (New Delhi: Oxford University Press, 1998). For discussions on colonial constructions of masculinity, see the chapters "The Creation of Difference" and "The Ordering of Difference" in Thomas Metcalf, *Ideologies of the Raj* (Cambridge: Cambridge University Press, 1998), 66–159; and Mrinalini Sinha, *Colonial Masculinity: The 'Manly Englishman' and the 'Effeminate Bengali' in the Late Nineteenth Century* (Manchester: Manchester University Press, 1995).

2 Tales of a city

Writing colonial Calcutta

Calcutta emerged as the principal urban centre of Britain's Indian empire and the capital of British India during the eighteenth century.[1] By the nineteenth century, the city had developed into a thriving commercial hub, and served as a crucial node in the global imperial political economy. It was an important port: a transit point for raw materials, labour for overseas colonies (often in the form of indentured Indian workers), and manufactured products for distribution within India. Its importance to the British imperial project in the opening years of the twentieth century was such that it was perfectly reasonable to suggest that, with "the exception of London, no city in the great British Empire can be compared to Calcutta in point of size, beauty, and commercial and political interest."[2] One of the earliest examples of colonial urbanization in South Asia, it was wrought in the crucible of capitalist modernity inaugurated by the British, and was firmly associated with their presence in the Indian subcontinent. Calcutta was one of South Asia's first "modern" cities. But it was not just a *new* urban space; it was also a new *kind* of city of South Asia that was distinct – in structure, look, and feel – from older, pre-colonial metropolitan centres such as Delhi, Agra, or Benares. Through the eighteenth century, and especially the nineteenth, Calcutta attracted numerous visitors from within South Asia, and beyond. To visit Calcutta was to encounter, and experience, modernity – and its attendant unevenness – spatialized in all its glory. Many of these visitors, some of them notable authors, wrote about the premier city of British India.

I look at some of these writings from the nineteenth and the earlier half of the twentieth century in this chapter to see how they inscribe the city – and often, its built environment – with widely differing meanings. My examination of the diachronic transformation of the city's representations will be anchored in constellatory readings of the imagined urban geographies of Calcutta from two moments of its history. For the first moment, from the early decades of the nineteenth century, I will examine poetic depictions of the city by a colonial official and part-time poet, James Atkinson, alongside the pre-eminent Urdu and Persian poet of the late-Mughal era Mirza Asadullah Baig – better known in South Asia as Mirza Ghalib – and a conservative Bengali intellectual, Bhabanicharan Bandyopadhyay. For the next moment, from the 1860s to the mid-twentieth century, I will constellate a range of literary and non-literary texts to look at their contested depiction of urban space. In this, I will take up two of the most important literary figures of British India, Rudyard Kipling and Rabindranath Tagore. In my readings I show the authors negotiating their experiences of Calcutta's contradictory space besides illustrating the contested ways of perceiving, conceiving, and being in the colonial city. My examination of these different and divergent representations of colonial Calcutta showcases the expression, and mutual contestation, of colonial and counter-colonial spatial desire at a local site. Moreover, they annotate, in a specific non-western context, Lefebvre's observation that the urban form, that is, any city anywhere in the world, forms an *oeuvre* that is developed not just by material goods and objects but also, and just as crucially, through the "production of knowledge, culture, works of art, and civilization."[3] And furthermore, this story of contestation showcases, in a scaled-down version, the larger contestation over the British Indian colony between the British and the Indians.

The making of a colonial metropolis

Calcutta's tryst with the English began in 1690 when the Mughal emperor Aurangzeb granted Job Charnock, an agent of the English East India Company, permission to carry on trade in Bengal. The Company was also allowed to establish a trading outpost in the village of Sutanuti, on the eastern bank of the western distributary of the Ganges (which the British named Hoogly) against the annual payment of 3,000 rupees. By 1695, Sutanuti became the seat of the Company's Bengal trade, and it purchased the rights to the taxes of villages of Sutanuti, Gobindapur, and Dihi Kolikata from the Sabarna Roy Choudhury family of landlords in 1698. This set the stage for the English transformation of the space into the city of Calcutta. The Company began building a fort on this land though it had no legal title to it. Fort William, named after William III of Britain, was completed in 1703. A hospital was built in 1709, and Calcutta's first church, St. John's, in 1716. Calcutta's 'White' Town, called "sahebpara" [সাহেবপাড়া] or "European quarters" in Bengali, was beginning to take shape.

In 1717, the Mughal emperor Farukh Siyar granted the British permission to acquire rights to the 38 villages surrounding the British settlement. While the British acquired the majority of the land rights, independent native zamindars continued to exist until 1757, "remaining outside the authority of the Company and Fort William."[4] In 1742, after a series of attacks by the Maratha raiders on a trading outpost in Howrah (across the river from Calcutta), the Europeans and Indians co-operated to excavate a 3-mile ditch north of the fort that was called the Maratha Ditch. The Company also began extending its fortifications in Calcutta, which the new Nawab of Bengal, Siraj-ud-daulah, saw as a challenge to his authority. In 1756, Siraj attacked and defeated the British garrisons at Fort William destroying much of the fort in the process.[5] The Company's forces, under Robert Clive, retook Fort William later that year. Subsequently, Clive conspired with some local merchants headed by Jagat Seth to replace Siraj with his disaffected general Mir Jafar. In 1757, the Company's forces defeated Siraj's army at the Battle of Plassey, which was largely achieved because a significant section of the Nawab's troops under Jafar stood aloof from the fighting. Jafar was installed as the Company's puppet Nawab who rid the area within the Maratha Ditch of all the independent zamindars. He also granted the Company 600 yards of land outside the ditch and rented them land in the south for an annual sum of 225,000 rupees.[6] The sacking of Fort William prompted the British to build a new fort at the site of the Gobindapur village, south of the old fort. For this, the residents of the village had to be removed. The 1876 census of Calcutta recalls this, noting that "many thousand huts [were] thrown into the holes from whence they had been taken, to form roads and an esplanade."[7] Prominent native landlords were compensated and resettled in other parts of the city, mainly in the north. The lower classes also moved to these neighbourhoods, giving shape to the 'Black' Town or native quarter that had already begun to form in the northern areas of Calcutta. The landlord families would shape the space of the 'Black' Town by re-enacting the socio-spatial practices of

> the little *rajyas* [fiefdoms] of the earlier period . . . where a kind of urbanism, at a level other than that of commercial cities or great politico-military centres, persisted through the centuries. . . . The comprador purchased land and settled it with tenants . . . [and] distribute[d] patronage on an elaborate scale . . . to acquire prestige and status.[8]

This resulted in the development of bazaars and hutments in the 'Black' Town where artisans and labourers lived and provided household labour to the elite families. These localities "became replicas of old landed estates owned by feudal chieftains, their contours changing in response to urban needs."[9] Calcutta's 'Black' Town, over time, became a city of palatial buildings of the elite Bengalis surrounded by bazaars and slums of hutments.[10] The construction of the new fort, also called Fort William, began in 1758 and was completed in 1781 at a cost of approximately 2 million pounds.[11]

The English also cleared the forests around the site of the fort resulting in the creation of a mammoth park at the heart of Calcutta that came to be known as the "maidan." In his history of Calcutta, James Rainey reminisces that the St. Paul's Cathedral, on the edge of the maidan, was once

> a good cover for tigers, and Warren Hastings [the first Governor-General of the English East India Company] hunted them on elephants there. . . . The maidan was a capital find for wild boars, and pigstickers of those days could indulge in this favorite pursuit where the gallant Volunteer Lancers now parade, and where fair equestrians now fearlessly roam.[12]

The maidan was an attempt at recreating a Hyde Park in colonial Calcutta, but it also provided the cannons mounted on the new Fort William an unrestricted view and command of the city and the river.[13] As the eighteenth century progressed, Calcutta's 'White' Town developed into a networked space that connected the administrative areas (centred around the old fort area) with the port, the commercial areas of offices, banks, warehouses, and wholesale markets, the residential neighbourhoods of mansions around the business district, as well as the garden houses in its outskirts. In 1773, the British Parliament passed the Regulating Act that overhauled the East India Company administration, elevating the Governor of Bengal (at that time, Warren Hastings) to the position of Governor-General, and placed the presidencies of Madras and Bombay under the control of Bengal. This made Calcutta the principal political centre of British India. In 1799 Governor-General Richard Wellesley decided to build a residence for himself, and his successors. Completed in 1803 and designed by Charles Wyatt on the lines of the Curzons' family mansion of Kedleston Hall in Derbyshire, Calcutta's Government House followed a neoclassical style with distinct Baroque overtones. It comprised a central core with four radiating wings and was based on a plan that allowed for a great deal of natural ventilation while also permitting its residents "to gather the view and breeze of the *maidan*."[14] On the south was a portico surmounted by a colonnaded verandah with a dome above. A balustraded wall surrounded the entire compound with a grand arched gateway at each of the four cardinal points, and the figure of a lion mounted on it. George Curzon, whose ancestral home served as the model, occupied the Government House as British India's viceroy in 1899. He would call it "the finest [residence] . . . occupied by the representative of any Sovereign or Government in the world."[15] In addition to the British and the native Bengalis, Calcutta attracted a substantial number of people from other communities. It became home to Marwaris from northwestern India, Odias from the southeast, Urdu and Hindi speakers from the northern Indo-Gangetic plains, as well as communities of artisans from other parts of Bengal.[16] Calcutta also had a significant presence of Armenian and Jewish merchants, in addition to the Portuguese, French, Dutch, Chinese, and Scottish businessmen and traders. These communities, as well as Eurasians, Muslim service groups, and the poor whites, settled around the edges of the European quarters to form the 'Grey' Town, a "heterogeneous intermediate zone" that met the diverse service requirements of the city.[17] This city's cosmopolitan character would continue to grow, and by 1901 its Bengali speakers were a slender majority, just above 51 per cent of the city's population.[18]

The British landscaping of Calcutta's 'White' Town continued apace through the late-nineteenth and twentieth centuries. The Calcutta High Court was built in 1862, on the model of the Stadt-Haus of Ypres; the General Post Office in 1864, its dome reminiscent of London's St. Paul's Cathedral; in 1878, the Indian Museum moved to a building constructed along neoclassical lines; in 1889, a Corinthian façade was added to the Writers' Building, seat of the colonial bureaucracy. After Victoria's death in 1901, the viceroy Curzon sanctioned constructing a "National Valhalla of British Indian worthies" in the memory of the late empress.[19] This took shape between 1906 and 1921 and became the Victoria Memorial Hall. Attempting a fusion of Mughal and Baroque architectural styles and built completely in white marble, it expresses the colonizers' desire to inscribe the native landscape with a British version of the Taj Mahal. By the nineteenth century, Calcutta was firmly associated with English presence in South Asia.

Three around Calcutta

Visitors to Calcutta's 'White' Town found its broad streets and avenues, its parks, as well as the stately neoclassical looks (sometimes with a touch of Baroque) of its business and administrative districts, and the grand residential mansions along the main streets, distinctly impressive. Over time, the city became the subject of a number of literary and visual representations. By the early-nineteenth century, Calcutta – or, more precisely, the 'White' Town – had earned itself the moniker "City of Palaces." This phrase can be traced to James Atkinson's 1824 poem by that name, although Kate Telscher notes that "descriptions of Calcutta's 'palaces' date from the 1780s."[20]

Atkinson's poem begins with a reference to past empires, before claiming that Calcutta "with its tombs/And dazzling splendors, towering peerlessly" was the porch of India, the blooming "garden of Hesperides" of the British.[21] Atkinson imagines India as a blooming garden, the description suffused with a sense of cornucopian plenitude associated with colonial possessions. Atkinson terms Calcutta the "porch" of the garden, that is, the man-made vantage from which the "natural" prospect of India is viewed. As such, the opening lines underline Calcutta's centrality to the British colonial mission in South Asia. A few stanzas later, Atkinson describes his moment of arrival in, and his experience of viewing, the city. Gazing at Calcutta's panorama from the river at the moment of arrival would be a recurring trope in the depictions of the city through the eighteenth and nineteenth centuries, all of them unambiguously adulatory.[22] Atkinson signals his reliance on this trope when he writes:

> I stood a wandering stranger at the *Ghaut*,
> And, gazing round, beheld the pomp of spires
> And palaces, to view like magic brought;
> All glittering in the sun-beam . . .[23]

The "spires" and "palaces" signal the stature of the urban space besides underlining the ocular of the poet's first encounter with the city. This visuality of urban experience is further complemented by the choice of words such as "gazing," "beheld," "pomp," "view," and "glittering." The person viewing Calcutta is presented as someone who is "like magic brought" to the city. Here, the claim of "magic" works in a number of ways. It gives a hyperbolic touch to the adulatory tone of Atkinson's poem once again gesturing towards the stature of Calcutta as a premier political and economic space of South Asia. But "magic" also signifies the unreality of the city in the sense of its *otherworldliness* – it marks the city out as a space that is singular, different from both the countryside as well as from other cities and towns in British India. And the invocation of magic, needless to say, occludes the material processes that undergird the British colonial project in South Asia. Calcutta for Atkinson remains, however, not a city in itself but a version of London. He calls Calcutta "a little London in Bengal," his simultaneous use of "London" as well as its diminution, both crucial for understanding the ideological production of the colonial city as secondary.[24] The city is "A microcosm; loose, and yet compact;/A smug epitome, a capital" and most certainly a product of the British colonial venture that is "formed like island on the main/ Amidst a sea of pagans, to exact/Allegiance from their millions. . . . For intellect hath power, to bind as with a chain."[25] The stress on Calcutta as an imperial possession continues throughout the poem and extends on to the other crucial marker of the city's space, the river. Addressing the river, the poet notes that "The Brahmins called thee theirs for ages . . . now navies ride/ Upon thy waters, – Strangers now have sway,/ And various nations throng to bear thy wealth away."[26] Though undoubtedly self-reflexive in its awareness of the looting of its colonial possessions ("various nations throng to bear thy wealth away"), the line also articulates a self-assurance and confidence in the British military superiority (the "navies" riding on the Ganges), and the inviolability of British rule in India ("Strangers now have sway").

Another aspect of the *City of Palaces* also merits attention. Though an early-nineteenth-century poem – it was published in Calcutta in 1824 – it, nevertheless, displays traits of Augustan literary creations reminiscent of the poems by John Dryden, Alexander Pope, and Samuel Johnson from about a century ago. This is evident in its constant deployment of personifications such as Nature, Victory, Fashion, Desolation, and Death throughout the poem. In addition, the poem invokes figures from Hellenic history and myths that include Diogenes, Antisthenes, Socrates, the Bacchanals, the "Syrens of the east," and "Heraclean Zeuxis" besides Pallas Athena, Jove, and Venus ("Love's fair queen").[27] The neoclassical traits of seventeenth- and eighteenth-century English literary creation – themselves a curious mix of cultural features of contemporaneous Britain and figures of Greco-Roman myth and history – in Atkinson's lines are rendered even more peculiar, not least because the Greek gods and notables in Atkinson's poetical universe co-exist with a variety of native South Asian markers such as "Brahmins," "Moonshees," and "nautch-girls," as well as the "Ganges" and the "hookah." The poet's use of the neoclassical style for depicting the city as well as what I call its formal peculiarity are actually both quite illuminating. This is because the poem, at the level of stylistics and form, seems to be registering the contradiction that is the hallmark of Calcutta's urban space. Atkinson's use of the neoclassical style for the poetic rendering of the city appears to draw sustenance from the neoclassical lay of Calcutta's 'White' Town. The European quarters of the city were a heartland of neoclassical architecture that were surrounded by neighbourhoods where the Indians lived, which were made up, overwhelmingly, of thatched huts.[28] This accorded the space of Calcutta with a contradictory character that was further accentuated during the day when native workers and clerks inhabited the neoclassical buildings in the city's business and commercial districts. It is this spatial contradiction of the urban space that the poem registers and mediates in the style and form. The peculiarity of Atkinson's poem is thus the result of a successful construction of verisimilitude as regards Calcutta's contradictory urban space.

Calcutta also attracted visitors from the late-Mughal court who often came to the city to petition and conduct business with East India Company officials. In

1828, four years after the publication of Atkinson's poem, the poet Mirza Asadullah Baig – better known as Mirza Ghalib – travelled from Delhi to Calcutta. Ghalib was in the city to petition the British for a pension that was due to him from his uncle's estate. While his mission remained unaccomplished, Ghalib returned to Delhi charmed by Calcutta's beautiful women and mangoes. He wrote about the colonial city in poems composed in Persian and Urdu as well as in several letters to his friends. Speaking about Calcutta in one of these letters, he writes:

> By God, had I not been a family man, with regard for the honour of my wife and children, I would have cut myself free and made my way there. There I would have lived till I died, in that heavenly city, free from all cares.[29]

But it is in another poem that Ghalib draws out a more evocative image of the colonial city drawing on the *shahrashub* or *shahrangiz* ("city disturber") tradition of pre-modern Persian poetry.

The *shahrashub* is more of a topos than a genre, and was originally a designation for a beautiful beloved in a lyric poem or a short bawdy lyric addressed to a young boy engaged in a trade or craft who coquettishly offers his wares to the love-struck poet.[30] As Sunil Sharma notes in his survey of the *shahrashub*, the rise of major urban centres of Persianate culture from the fifteenth century transformed this kind of poem into a "unified work specifically written for a city and ruler, [with these providing] information on professions and crafts in various cities and times in history."[31] In South Asia, this poetic form develops in the Nizam Shahi and Adil Shahi courts at Ahmadnagar and Bijapur of the Deccan, with the first Indo-Persian poet of note being the Persian émigré Nuruddin Muhammad Zuhuri. In north India, the form contributed to the large corpus of Mughal poetry about buildings and gardens, especially through the work of Abu Talib Kalim Kashani, Shah Jahan's poet laureate. The *shahrashub* would be rendered in both prose and in poetry, where the latter typically took the form of a *masnavi*, the premier Indo-Persian narrative genre before the advent of the novel.[32] A key feature of the *shahrashub*, in both poetry and prose, was the catalogue device that Sharma suggests was an "indirect precursor to the ethnographic surveys of the colonial period that mapped out the complex fabric of Indian society into a detailed typology according to castes and communities."[33] With the waning of Mughal power through the eighteenth and nineteenth centuries, poets redefined the function of the city poem: "the exuberant city poem of Persian with *shahrashub* elements became the *shahrashob* (the disturbed city), a lament for a declining city in classical Urdu poetry."[34] Poets such as Mir Taqi Mir and Nazir Akbarbadi wrote *shahrashubs* in a satirical mode, providing a bleak picture of urban society amidst the collapse of commerce and social hierarchies.

In his poem on Calcutta, Ghalib relies on this long and pre-modern literary tradition even as he repurposes it for depicting the modern colonial city. The poem, structured as a short conversation between Ghalib and his guiding angel, has him commenting on Delhi, Benares, and Azimabad, before moving on to Calcutta. Delhi is the "soul/and the world its body"; Benares a "sweet beloved absorbed/in plucking flowers" while Azimabad is more "colourful than/ a garden's air."[35] Of the colonial city, Ghalib writes:

> I asked about Calcutta and he said:
> "It should be called the Eighth Clime."
>
> I said, "Do people come together in it?"
> He said, "From every land and of every trade."
>
> I said, "What business turns a profit here?"
> He said, "Starting in fright from all that is."
>
> I said, "What work should one do here?"
> He said, "Quit poetry and writing."
>
> I said, "Who are these lunar beauties?"
> He said, "The beauties of the land of London."
>
> I said, "Might they have hearts?"
> He said, "They do but of iron."[36]

Ghalib draws on the familiar devices of the *shahrashub* by laying out the people and professions of the modern city. However, the poem is also an innovative rendering of the topos where Calcutta, the new metropolis, is presented as an incarnation of the marvellous city of the beloved found in the Indo-Persian literary tradition. And while the city is undoubtedly a spectacular urban space for Ghalib, it is not beyond reproach. It is primarily a city of commerce where the profitable profession is to be afraid "from all that is"; and where one can do any manner of work but poetry and writing. And though Calcutta is a city peopled with fair creatures, namely, the "lunar" beauties of London, they have hearts made of iron. It is significant that Ghalib celebrates, and critiques, the colonial city simultaneously. This allows him to innovate the pre-modern *shahrashub*, fashioning it into an ambivalent mode of address for the modern city.

Not all visitors to the colonial city engaged with it in an ambivalent manner. The Bengali satirical tract *Kalikata Kamalalay* [কলিকাতা কমলালয়], which appeared in 1823, a year before Atkinson's poem, takes a more critical view of Calcutta and the Calcuttans.[37] Bhabanicharan Bandyopadhyay, the author, displays a remarkable degree of unease at the city, not least for the rudeness of the city's residents towards people from the countryside. This is not surprising as the "modernity" of Calcutta was often contrasted with the "backwardness" of the countryside and the rustic bumpkins – gñeyo [গেঁয়ো] is the Bengali word for it – who lived there. A collection of four short parts – its author calls the parts "waves" – it is ostensibly designed to introduce the country folk to the city. The author claims that when people:

> come to Calcutta from villages and other smaller towns and cities . . . they sit silently among the Calcuttans in a rustic manner . . . when a large number

> of citizens (of Calcutta) sit together and argue . . . [and] a man from a village offers even a correct reply . . . the noblemen of the city would not agree to it . . . [They tell him] you are a rustic . . . you need not bother about these things.[38]

He therefore justifies his project of "giving a crude description of the great city of Calcutta . . . [so the reader can] realize easily the behavior, customs and the art of speaking prevalent [t]here."[39] The author's primary concern in *Kalikata Kamalalay* is to critique the lax religious Hindu codes in the city. For this, he delineates an experiential landscape of the city that revolves around money, ceremony, and commerce. It is a melting pot of people on one hand, and the site for the 'corruption' of Hindu mores on the other. He also underlines the 'distortion' of the Bengali language in Calcutta, that is, the entry of a host of loan words from English, Portuguese, Urdu, Arabic, and Persian into what was traditionally a Sanskrit-based vocabulary. Despite his articulated conservatism, some of his observations are illuminating. Describing the city as a sea, he writes:

> Calcutta is full of wealth like unfathomable water of the sea which is undrinkable . . . [During] a big ceremony money flows like water in different directions. Different kinds of money transaction make the money circulate constantly like the rivers constantly carrying water to the sea.[40]

By constructing an equivalence between the circulation of money and the flow of water, Bhabanicharan describes, quite accurately, the flow of capital into the nineteenth-century city besides outlining its location as a node in the global circulation of capital. This is further underlined by the author's equation of Calcutta with the sea, as it points towards the city as the destination, and repository, of capital from the hinterland, thereby highlighting once again, its role in the expropriation of the countryside.

During the high imperial era of the late-nineteenth century, Calcutta was considered a city of global importance. Texts from the "great cities of the world" genre regularly listed Calcutta along with other major cities from across the world. An 1886 text opens the description of Calcutta with these lines:

> WE approach the "City of Palaces" by the river. . . . Everywhere there are tokens that we are drawing near to a great city. As we proceed up the river . . . here and there are detached bungalows, factory chimneys, fishing and trading boats; farther on still, merchantmen lie anchored in the stream, and steamers pass and re-pass, filled with passengers. At the beautiful suburb of Garden Reach, the eye, weary of the low swampy banks and interminable jungle, is refreshed with a profusion of elegant and stately residences in the midst of masses of luxuriant verdure. In front of the ornamental villas, grassy lawns and rich flower-beds slope down to the water's edge; the river, as Calcutta is neared, is crowded with shipping – amongst it some of the finest vessels in the world; and by-and-by the usual landing-place in front of the Maidan is approached.[41]

The text here uses the trope of arrival, and in so doing contributes to its prevalence in literary depictions of the city. The opening section draws attention to what by then had become iconic about the Calcutta landscape: the busy port, the ornamental villas, stately residences, luxuriant verdure, and the maidan by name. This tenor continues in speaking of other aspects of the city's built environment:

> The main streets of Calcutta are broad and clean, with pleasant squares and avenues of trees delighting the eye with their refreshing greenness. . . . The business streets are lined with the shops of European traders, and mostly run directly down to the river, or along its banks. Here auction marts, warehouses, shops and offices, crowding carriers and porters, and bustling merchants may remind the Englishman of scenes nearer home.[42]

As a text that disseminated notions of cities from the world in the metropole, it also makes the reader see the city in a certain way, thereby reifying a fixed perspective: a specific way of seeing the city. It partakes in, and helps construct, the iconicity of the city, itself a function of Calcutta's centrality to the British colonial enterprise in Asia. By the mid-nineteenth century, the idea of Calcutta as a spectacular space and as the city of palaces would give way to another perspective on the city, just as reified and iconic. It would gradually come to be seen as an epitome of filth and squalor.

From palaces to pale faces

Around the mid-nineteenth century, the colonial state had begun to focus its attention to the state of Calcutta's civic infrastructure and found it decidedly wanting. Among the first to raise a stink was the President of the Bengal Sanitary Commission, John Strachey (recall, we have encountered his ideas about India in the Introduction). His comments on the city are preserved in a minute dated 5 March 1864 and are reproduced in the Commission's first annual report. It notes:

> The President of the Commission . . . stated that in the filthiest quarters of the filthiest town that he had ever seen in other parts of India or other countries, he had never seen anything to be compared with the filthiness of Calcutta . . . this was true, not only of the parts of the city inhabited by the poorer classes, but of the quarters filled with the houses of the richest and most influential portion of the native community . . . Mr. Strachey declared that the state of the capital of British India, one of the greatest and wealthiest cities in the world, to be a scandal and disgrace to a civilized government.[43]

Strachey's remarks on the city as a site of filth and squalor inaugurates another way of seeing, and showing, Calcutta. This perspective would become the normative way of perceiving the city for the British as the nineteenth century progressed, displacing the older ways of writing about it. And as we shall see, Strachey's precise words would be cited and re-cited as the idea of Calcutta as a filthy city gained wider currency.

In 1869, a few years after the first annual report of the Bengal Sanitary Commission was published, Calcutta's Sanitary Commissioner David Smith authored a report on the city's drainage and conservancy where he presents, as he terms it, a "graphic picture" of the city's sanitary condition. Smith's report includes an extended extract from Strachey's minute with some its sections italicized:

> The state even of the Southern Division of the Town, which contains the fine houses of the principal European inhabitants, is often most offensive and objectionable, while, with regard to the Northern or Native Division of Calcutta, which contain some hundred thousand people, *it is no figure of speech, but the simple truth, to say that no language can adequately describe its abominations. In the filthiest quarters of the filthiest towns that I have seen in other parts of India, or in other countries, I have never seen anything which can be, for a moment, compared with the filthiness of Calcutta.* This is true, not merely of the interior portions of the town or of the bye-ways and places, inhabited by the poorer classes but it is true of the principal thoroughfares and of the quarters filled with the houses of the richest and most influential portion of the Native community. If a plain unvarnished description of the streets of the Northern Division of Calcutta, bordered by their horrible open drains, in which almost all the filth of the City stagnates and putrefies, were given to the people of England, I believe that they would consider the account altogether incredible . . . *the state of the Capital of British India, one of the greatest and wealthiest Cities in the world, is a scandal and a disgrace to civilized Government.*[44]

Smith's and Strachey's reports both speak of a crisis in public amenities that was quite the daily reality of Calcutta in the late-nineteenth century. Both mention the need for better drains and sewers, and the urgent need for upgrading the infrastructure of both the 'White' and the 'Black' Towns of the city. This prevalence of filth everywhere, even if more prevalent in the 'Black' Town, was in a perverse way mitigating the contradictory character of the Calcutta landscape. But these comments are also significant as they highlight the failures of the colonial state to effectively manage and administer the city.

The British had been governing the urban space of Calcutta since the grant of the *diwani* of Bengal to the English East India Company in 1765, which made them the titular revenue minister of the Mughal monarch and allowed them to collect taxes from the Bengal province. This, however, did not imply a legal or moral obligation on the Company to render civic services to the city, and none were provided. The demand for municipal services grew after 1773 when Calcutta became the centre of British operations in India, and statutory civic services began only in 1794. From 1785, Calcutta was divided into 31 *thanas* [থানা] or "wards" for municipal and police purposes. To remove garbage, the colonial administrators assigned four bullock carts per *thana* in the 'White' Town, and two for every *thana* in the native part of the city.[45] By 1802, the city had 85 bullock carts, with two municipal depots to house the bullocks and the carts. Calcutta's sewerage system also developed late. In 1695, a trench was dug around the Sutanuti factory and in 1710, another, deeper trench was created "to separate the British settlement from the native quarters and keep the former dry and wholesome."[46] The Maratha Ditch served as the grand drainage outlet of Calcutta before it was filled up between 1799 and 1801. The city had surface drains everywhere by 1857 and, in 1859, a plan for underground drainage was sanctioned and completed in the next 15 years. The year 1864, the year of Strachey's illuminating comments on Calcutta's infrastructure, was particularly important in the city's civic history. The Calcutta municipal administration acquired land that year at Dhapa, in the eastern fringes of the city adjoining the salt marshes, to create the "municipal square mile," the giant garbage dump of the city. Until then, Calcutta's human excreta – the evocative euphemism for which is "night soil" – were collected by the *mehtar* [মেথর] (manual scavengers) who were privately engaged for the task. This was taken to "night soil" depots and from there by bullock cart to the Night Soil Ghat on the banks of the river. From there, boats hired by the municipality carried it out on to the Hoogly and dumped it downstream at ebb tide, 200 tonnes of it daily. This practice was, however, rendered "more objectionable still by lax execution, the greater part of the filth thrown into the public drains under cover of darkness."[47] From 1864, the municipal administration arranged for trenching part of the human waste at Dhapa. A railway was built to carry the city's refuse to the municipal square mile in 1867, and more trenching sites were opened by 1896.

By the late-nineteenth century, the knowledge of Calcutta's dismal lack of civic infrastructure, as well as the gradual improvements in that area, were emerging out on to the larger public domain. Hodder's *Great Cities of the World*, which we have seen write approvingly about Calcutta's vista from the river, also had a section titled "Its Unhealthiness" that commented on the city having "little regard to sanitary arrangements [claiming instead that] some parts of it being below the level of the river [made] . . . drainage . . . a difficult problem to reformers of later date."[48] And while suggesting that sanitary conditions were improving in Calcutta, it would also use Strachey's words from the minute of 1864 to make its case:

> Of late years there have been vast sanitary improvements, and greatly they were needed. Only twenty years or so ago the president of a Sanitary Commission in Bengal wrote thus: "In the filthiest quarters of the filthiest towns that I have seen in other parts of India or in other countries, I have never seen anything which can be for a moment compared with the filthiness of Calcutta." There was then a lamentable absence of drainage; the streets were saturated with sewage, the air was filled with poisons, and in consequence dysenteric diseases were as common as catarrhs are in England. . . . Since then the sanitary state of the city has been improved from time to time and measures taken to modify the glaring contrasts which existed during the early part of the present century between the native town and the English

> quarter. . . . It is an old joke of residents who are complimented upon their "City of Palaces" to speak of it in return as the "City of Pale Faces."[49]

This description of the city's unhealthiness is striking. While it speaks accurately of the city's lack of civic infrastructure, it nevertheless fails to mention the British failure to build them up though they controlled Calcutta uninterruptedly since 1757. Instead, the lack of hygiene and filth are ascribed as features intrinsic to the land itself by suggesting, inaccurately, that the city was below the level of the river. And while Hodder's popular geography text quotes from Strachey's report to back its contention up, it omits the Sanitary Commissioner's own indictment of the city's British administrators ("a scandal and a disgrace to civilized Government"). Instead, the text undercuts its earlier presentation of Calcutta as a city of palaces by introducing a gentle dose of humour: by making a pun between "palaces" and "pale faces," and between "faces" and "faeces."

The text, moreover, deploys the space of the "Native" town to present Calcutta as a city *unlike* London or any other European city. The popular geography text writes that

> a large portion of the extensive area covered by Calcutta presents scenes such as have no counterpart in European cities. The native city, or Black Town . . . consists chiefly of narrow, dirty, and unpaved streets, crowded and stifling with dust and heat, with long rows of low brick buildings used as shops and warehouses, huts built of mud, or of cane and matting, wooden cottages raised on piles, all mixed in motley confusion.[50]

The text in making the claim that Calcutta's 'Black' Town had no counterpart in European cities seems, or chooses, to forget the existence of London's East End, the poor neighbourhood of the metropolitan city. There exists, of course, a robust corpus of works on the working class and urban poor of nineteenth-century England.[51] A look at some of their descriptive protocols necessarily undermines the claim to singularity for Calcutta's filth and squalor. Take, for instance, Reverend Andrew Mearns's *The Bitter Cry of Outcast London* from 1883 that purports to be, as its subtitle suggests "*An Inquiry into the Condition of the Abject Poor*." Mearns documents the residents of London's East End in a language that parallels the descriptions of not only the colonial government's reports and the popular geography text, but also that of Kipling's descriptions of Calcutta which I will discuss later. Take, for instance, this description:

> Few who will read these pages have any conception of what these pestilential human rookeries are, where tens of thousands are crowded together amidst horrors which call to mind what we have heard of the middle passage of the slave ship. To get into them you have to penetrate courts reeking with poisonous and malodorous gases arising from accumulations of sewage and refuse scattered in all directions and often flowing beneath your feet; courts, many

> of them which the sun never penetrates, which are never visited by a breath of fresh air, and which rarely know the virtues of a drop of cleansing water.[52]

The geography text clearly disavows this corpus of writing on London to set up Calcutta as an exceptional space. Yet, at the same time, it uses a similar descriptive and rhetorical mode that draws on the visual, olfactory, and tactile registers to structure an evocative anti-aesthetic description. In other words, it fashions a "slumming narrative" that was familiar to, and popular with, metropolitan readers in Britain to present the poverty of Calcutta, and its 'Black' Town in particular, as both singular and iconic.[53]

Reports on Calcutta in British-run English newspapers of British India also highlighted the contradictory nature of the city's landscape. Newspapers such as *Englishman* and *The Statesman* of Calcutta, as well as *The Pioneer* of the "up-country," that is, north Indian, town of Allahabad, significantly contributed to the increasing equation of the city with filth and squalor, further crystallizing this view. Further, the publisher Thacker, Spink and Company, publishing out of Calcutta, Bombay, and London, incorporated the pieces on Calcutta's 'Black' Town from *The Statesman* and *Englishman* in its 1890 edition of the *Thacker's Guidebook to Calcutta*. This necessarily provided the articles with a pan-Indian audience, as well as a readership in Britain, thereby disseminating widely these disparaging accounts of the city. Particularly important to note here is a change in the perception of the colonial city around the mid-nineteenth century. It is no more a "City of Palaces" or a "little London in Bengal" *alone* but *also* a "filthy city." To put this another way, the contradictory urban space of Calcutta had begun to inform its British representations. It was now *both* a city of splendour and squalor.

The contradictory character of Calcutta's urban landscape informed not only the content but also the form of some of these texts. The structure of *Thacker's Guidebook to Calcutta* is particularly illuminating in this regard. Divided into two sections, the first deals exclusively with the 'White' Town, outlining in detail the various colonial constructions of the city in addition to the history of the colonial buildings.[54] This section provides empirical details about specific sites – concrete spaces locatable on the city's map – that visitors could visit and see as material signs of British rule in the urban landscape. The second section, on the contrary, presents vignettes on the 'Black' Town published earlier in the *Englishman* and *The Statesman*.[55] Moreover, the titles of these pieces – "The Lepers and the Poor," "The Lunatic Asylums," and "The Municipal Square Mile" – suggest these spaces are beyond the purview of the city's visitors. In addition to mapping out in textual form the spatial contradiction of Calcutta, the guide also posits another kind of contradiction in the cityscape: between the rational space of progress marked by the physical form of the 'White' Town, and the deviant, backward space of the 'Black' Town that is presented in an affective register.

Noteworthy in *Thacker's Guidebook to Calcutta* is the invocation of "The Municipal Square Mile," the city's open garbage dump. Describing the space with more than a hint of irony, the text notes that this was a place that would "make the proverbial thousand distinctive smells of Cologne fade into insignificance."[56]

It describes the vegetation and animal life around the site while informing readers that it is bordered by human habitation. Describing the villagers as paddy cultivators and fisher-folk, it tells readers that the villagers "eat their food only after dark, owing to the swarm of flies which in the day time rendered even cooking a matter of difficulty."[57] Amidst this vision of intense degradation, the narrative holds out some hope of "improvement" for the elite residents of Calcutta. It claims that while the distinctive smell of Dhapa sometimes wafts towards the city – and as we will see subsequently, Kipling makes much of this issue while writing about Calcutta – the liquid sewage, though the breeding ground for malaria, does not actually flow into the Hoogly because of several locks that regulate its flow.[58] The articles from the Calcutta-based newspapers replicated in *Thacker's Guidebook to Calcutta*, then, underline the contradictory nature of the city landscape. It presents the 'White' Town as a city of stately buildings while also depicting the native spaces of the city, marked by filth and stench, in an affective register and as deviant. The articles appear to suggest that efforts were not only being made by the British administrators to mitigate the deviance and backwardness of the 'Black' Town, but that progress was being made in that direction. But if these texts hold up Calcutta as a contradictory space, Kipling's articles on Calcutta – which I will take up next – provides yet another view of the city. Invoking the "Big Calcutta Stink" as the great leveller, these articles would proclaim Calcutta as an infernal space where the native quarters of the city are simply the deeper circles of the same hell.

The clammy odour of blue slime

Kipling's vignettes on Calcutta were published as individual articles in *The Pioneer* between February and April 1888, and later anthologized as *The City of Dreadful Night* in 1890. A series of eight essays on the city, it narrates the author's experience of the 'White' Town before proceeding to introduce readers to the city's native quarters. The journey into the various areas of the 'Black' Town is presented as analogous to a descent into the various circles of hell (one of the essays is titled "Deeper and Deeper Still") with the members of the Calcutta Police playing Virgil to the author's Dante.

The City of Dreadful Night opens with a comparison between the urban landscapes of London and Calcutta. Kipling claims that anyone driving across the bridge over the Hoogly into the city could say that they had "left India behind [to enter] foreign parts."[59] In what develops into a play on un/familiarity, this is quickly rearticulated as "No, not wholly foreign. Say rather too familiar."[60] Owing to the visual similarity between London and Calcutta, the colonial city reminds Kipling of the "lost heritage" of its metropolitan counterpart: "the roar of the streets, the lights, the music, the pleasant places, the millions of their own kind, and a wilderness full of pretty, fresh-coloured English-women, theatres and restaurants."[61] Calcutta, however, is not London; it simply *reminds* him of the English city holding out "false hopes of some return" where the falsity of the colonial space is indicated through the imagery of smoke and foul smell.[62] In the Indian city, the "dense smoke hangs low, in the chill of the morning, over an ocean of roofs, and, as the city wakes, there goes up to the smoke a deep, full-throated boom of life and motion and humanity."[63] And while London is described with a degree of specificity using words such as "streets," "lights," "music," and significantly, "fresh-coloured English women," the smoke of Calcutta obscures any such descriptive effort becoming instead, an "ocean of roofs" housing an indistinguishable "humanity." In his desire to stress the *difference* between the colonial and metropolitan cities, Kipling omits any mention of the legendary London fog that appears repeatedly in its eighteenth- and nineteenth-century depictions. And he also appears unwilling to acknowledge the role of British jute mills in the Calcutta hinterland in producing the "dense smoke [that] hangs low" in the morning.[64]

The smoke inspires Kipling to disavow the faculty of vision altogether, for he claims that in India, the European "eye has lost its sense of proportion, the focus has contracted . . . and the mind has shrunk with the eye."[65] Consequently, he can mock the sentiment of seeing Calcutta and exclaiming "Why, this is London! This is the docks. This is Imperial."[66] The scepticism towards the faculty of vision underlines both its *fallibility* as well as its *inability* to grasp "reality" in this alien colonial space. This disavowal of vision, of course, has important consequences. The visual similarity between the landscapes of the metropolitan and colonial cities would destabilize the difference between the ruler and the ruled. Kipling's narrative strategy must therefore necessarily repudiate the faculty of vision, to retain the distinction between the two spaces, and uphold the rule of difference undergirding colonialism. Writing about the visual culture of nineteenth-century Britain, Kate Flint has noted that it was "characterised not just by the accelerated expansion of diverse opportunities for differing sorts of spectatorship, but [also] by a growing concern with the very practice of looking, and with the problematisation of that crucial instrument, the human eye."[67] Flint's contention of the Victorians' contradictory impulse towards vision and visual technologies acquires a more potent shade in the colonial context. On one hand, the visual register, through photography and, especially, its indexical nature, claimed a "stern fidelity" to truth.[68] Used extensively in colonial mapping and ethnography, it proclaimed the backwardness of India and Indians, and legitimized colonial rule.[69] On the other hand, the disavowal of vision and its attendant technologies allowed for the denial of spatio-temporal coevalness, and helped consolidate the rule of difference.[70]

With the visual register discarded, the narrative subsequently must rely on other sensory registers to structure its claim regarding the differential experiences of the spaces of London and Calcutta. In Kipling's narrative, the faculty of smell becomes the key to anchor the idea of spatial difference. Once again playing on similitude and difference, but now with the stenches of various British Indian cities vis-à-vis Calcutta, Kipling tells his readers that while Benares is "fouler in point of concentrated, pent-up muck, and there are local stenches in Peshawur . . . for diffuse, soul-sickening expansiveness, the reek of Calcutta beats both Benares and Peshawur."[71] Kipling then proceeds to take his readers on a guided olfactory tour of the Calcutta landscape. He writes:

> The Big Calcutta Stink . . . resembles the essence of corruption that has rotted for the second time – the clammy odour of blue slime. . . . There is no escape from it. It blows across the *maidan*; it comes in gusts into the corridors of the Great Eastern Hotel . . . the "Palaces of Chouringhi" carry it; it swirls round Bengal Club; it pours out of by-streets with sickening intensity, and the breeze of the morning is laden with it. It is first found, in spite of the fume of the engines, in Howrah Station. It seems worst in the little lanes of Lal Bazar . . . but it is nearly as bad opposite the Government House and in the Public Offices. . . . Six moderately pure mouthfuls of air can be drawn without offence. Then comes the seventh wave and the queasiness of an uncultured stomach.[72]

The olfactory mode in Kipling's description serves the important function of introducing the experiential register into the landscape form. While the urban landscape of London is seen and heard, Calcutta's is seen, heard, and crucially, *smelt.* In other words, it is articulated as *experienced* and becomes a *felt* reality. All the places mentioned in the extract above, significantly enough, are colonial buildings from the city's 'White' Town. The smell that suffuses the buildings makes them, and therefore the urban space these constitute, decidedly different from London's. It is this experiential register of the landscape that actually structures the charge of Calcutta as a differential urban space. Kipling of course does not explain, let alone justify, his assertions about Calcutta's "stink." Instead of situating it in the lack of the city's civic infrastructure – and the failure of the colonial state to provide them – he attributes it to the lack of "high-handed oppression." The stink is, willy-nilly, the fault of having natives on the board of management. He writes:

> If an up-country station holding three thousand troops and twenty civilians owned such as possession as Calcutta does, the Deputy Commissioner or the Cantonment Magistrate would have all the natives off the board of management or decently shovelled into the background until the mess was abated. They then might come on again and talk of "high-handed oppression" as much as they liked.[73]

His reference to the "board of management" is a possible allusion to the elective system that was introduced in Calcutta's municipal administration from 1876. From that year, two-thirds of the Municipal Commissioners were elected, while the rest – including the Chairman – were government nominees. Kipling's outburst clearly militates against any form of participatory government between the colonizer and the colonized.

Throughout the articles, Kipling provides his readers with a sensuous tour of Calcutta and its inhabitants: its sights, sounds, and smells. The Bengali, always the butt of Kipling's barbs in his fiction and non-fiction, is not spared in Calcutta either.[74] Kipling mocks the Bengali characters he meets, as in the case of the Bengali speaker in the Bengal Legislative Council chamber who quotes John Stuart Mill to argue for giving prominence to the "popular element in the electoral body."[75] And in addition to the natives, Kipling finds Calcutta populated by foreign types. Near the Calcutta Port, he finds:

> Malay-Hindus, Hindu-Malay-whites, Burmese, Burma-whites, Burma-native-whites, Italians with gold earrings and a thirst for gambling, Yankees of all the States with Mulattoes and pure buck-niggers, red and rough Danes, Cingalese, Cornish boys . . . tun-bellied Germans, Cockney mates . . . unmistakable "Tommies" who have tumbled into seafaring life . . . cockatoo-tufted Welshmen . . . broken-down loafers, grey-headed, penniless, and pitiful, swaggering boys, and very quiet men with gashes and cuts on their faces.[76]

Kipling's description gestures towards the multi-ethnic character of the city with these social groups – the Bengali, the Englishman, as well as the foreign "types" Kipling describes for us – brought together by the forces of colonial globalization. But the city is not quite the melting pot of humanity for him. Instead, one can detect in Kipling an enumerative modality that is typically associated with the practices of colonial anthropology. It is hardly surprising, then, that he thinks of the city as an "ethnological museum where all the specimens [we]re playing comedies and tragedies."[77] Notice his deployment of the terms "museum," "tragedies," and "comedies." For the English author the city is less the seat of British authority or a site of history. Rather, Calcutta appears to exist outside of history and as a museum or a playhouse stage. Not only does this allow him to conceal the place of Calcutta as a vital node in the global political economy of colonial capitalism, it also allows him to recode the city as an oddity and singular in its difference from normative spaces such as London. But even as he describes Calcutta, and reduces its stature by caricaturing it, what does come across is his sense of bewilderment and an almost desperate desire to "make sense" of one of the first and truly global cities of South Asia. And this perspective on Calcutta aligns Kipling, curiously enough, to Bhabanicharan Bandyopadhyay. But unlike Bhabanicharan, whose sense of astonishment was owing to Calcutta's wealth and the Calcuttans' rudeness, Kipling here successfully conceals his perplexity by recoding it with racial and imperial arrogance.

As the narrative progresses, Kipling takes his readers into the 'Black' Town, the "darker portion . . . [that] does not look an inviting place to dive into at night."[78] Invoking a testimonial narrative stance, the author claims to unveil for his readers a "part of the world [of Calcutta that is] . . . shut to Europeans – absolutely."[79] His escort from the Calcutta Police tells him that they are in "St. John's Wood of Calcutta – for the rich Babus."[80] This provokes in the author a return to the trope of the stench. He writes that this part of the 'Black' Town is filled with

> mysterious, conspiring tenements as Dickens would have loved . . . [with] no breath of breeze . . . and the air . . . perceptibly warmer [than the 'White' Town]. . . . There is . . . utmost niggardliness in the spacing of what, for want of a better name, we must call the streets. The air is heavy with a faint, sour

> stench – the essence of long-neglected abominations – and it cannot escape from among the tall, three-storied houses.[81]

Accompanying the police in this dark side of the Calcutta landscape, Kipling reaches the palatial mansion of an Indian woman who "stands revealed, blazing – literally blazing – with jewelery [sic] from head to foot."[82] He spends a considerable amount of space extolling the beauty of, what he calls, the "Dainty Iniquity." He describes – and tries to estimate the value of – her ornaments besides providing an account of her house, furniture, and decorations.[83] Kipling's ethnographic eye records, and having recorded moves on once his "Dainty Iniquity" departs with her paramour ("Fat Vice" is Kipling's appellation for the latter).

The colonial Dante now ventures deeper and deeper into the circles of his tropical hell, and into Calcutta's red-light district. About this more profound space of Calcutta's darkness, Kipling writes: "On one side are houses – gaunt and dark, naked and devoid of furniture; on the other, low, mean stalls, lighted, and with shamelessly open doors, where women stand and mutter and whisper one to another."[84] And here he encounters a prostitute, Mrs. D, the Eurasian widow of "a soldier of the Queen, [and the] mother of seven children."[85] Mrs. D, he writes, "has stooped to this common foulness in the face of the city . . . [and] has offended against the white race."[86] The members of the Calcutta Police who accompany Kipling do not share his sense of indignation however. They tell him: "You're from up-country, and of course you don't understand. There are any amount of that lot in the city."[87] Note the policeman describing Kipling as someone from "up-country" who doesn't quite understand the ways of Calcutta. This is illuminating: not only is the policeman, a local resident, dismissing the moral and moralizing concerns of the author, a petulant outsider, but he also appears to scorn Kipling's small-town provenance ("up-country"), making a distinction between the north Indian town of Allahabad and the metropolis of Calcutta. Kipling here seems to be experiencing the curtness and disdain that Bhabanicharan Bandyopadhyay had noted in Calcuttans towards visitors from smaller towns and villages almost half a century earlier.

His companion's derision, however, does not stop Kipling from presenting Mrs. D as an incarnation of moral degeneracy. Her offence is twofold: she is Eurasian, that is, born of Indian and English parents; she is also depicted as a prostitute who entertains Indians as customers. Kipling champions stricter racial – and spatial – boundaries, reminiscing about the "good old days" when administrators "deported him or her that misbehaved grossly, and the white man preserved his *izzat* [honour]. He may have been a ruffian, but he was a ruffian on a large scale."[88] While Mrs. D allows Kipling to narrate the darker spaces of Calcutta in a sensational vein, her presence in the narrative nevertheless threatens to undermine its neatly racialized worldview.[89] His account thus seeks to contain the figure of Mrs. D, not least by underlining her Eurasian provenance and therefore invoking the widely held belief that they, owing to their racially mixed provenance, were morally inferior. Mrs. D, her profession (prostitution), and her choice of customers (Indians) all become symptomatic of the racial miscegenation she embodies besides serving as a warning of sorts against the breaking of racial boundaries. The story of Mrs. D – presented, at once, as repugnant and lurid – would be a "sensational read" for Kipling's audience, and appeal to the normative tastes of the English readership in India. By situating these figures within Calcutta, in particular the 'Black' Town, Kipling also appears to underline how the physically "filthy" city animates, and is in turn animated by, these figures of "moral filth" such as Mrs. D.

Calcutta's urban space is truly irredeemable for Kipling, and he does not even spare the English dead who lie buried in the city. His diatribe extends to the Old Park Street Cemetery, one of the earliest non-Church cemeteries in the city that was used from 1767 to the 1830s. Commenting on its built environment, Kipling writes:

> The eye is ready to swear that it is as old as Herculaneum and Pompeii. The tombs are small houses. It is as though we walked down the streets of a town, so tall are they and so closely do they stand – a town shrivelled by fire, and scarred by frost and siege.[90]

Kipling's invocation of Herculaneum and Pompeii once again posits Calcutta as museum and relic of the past, in effect an oddity. He also demonstrates scant regard for the structure of the tombs and mausolea in the cemetery, constructed with bricks and lime, that combine Gothic elements with the Indo-Saracenic style popular with the British in India.[91] He derides the built form of the tombs further, noting:

> They must have been afraid of their friends rising up before the due time that they weighted them with such cruel mounds of masonry. Strong man, weak woman, or somebody's infant son "aged fifteen months" – it is all the same. For each the squat obelisk, the defaced classic temple, the cellaret of chunam [lime], or the candlestick of brickwork – the heavy slab, the rust-eaten railings, the whopper-jawed cherubs, and the apoplectic angels. Men were rich in those days and could afford to put a hundred cubic feet of masonry into the grave of even so humble a person as "Jno Clements Captain of the Country Service 1820."[92]

Kipling appears intent on demonstrating the derivative nature of its built form. By focusing on the *secondary* nature of the European architectural models in colonial India ("obelisks" and "classic temple") and the building materials ("chunam" and "brickwork" instead of, say, marble), he projects the *second* city of the British Empire as a bad imitation of the original, a *secondary* city. In mocking the "Captain of the Country Service," Kipling appears to suggest that the colonial officials buried within its premises were the refuse of history, and the cemetery, little more than its dustbin.

Kipling directs his barbs, in particular, towards a particular tomb: the "big and stately tomb sacred to Lucia who died in 1776 AD aged 23."[93] In a virtuoso display of high imperial arrogance and sarcasm that does not even spare the British East India Company from ridicule, he writes:

> What pot-bellied East Indiaman brought the "virtuous maid" up the river, and did Lucia "make her bargain," as the cant of those times went, on the first, second, or third day after her arrival? Or did she . . . give a spinsters' ball as a last trial – following the custom of the country? No. She was a fair Kentish maiden, sent out, at a cost of five hundred pounds . . . under the captain's charge, to wed the man of her choice, and *he* knew Clive well . . . and talked to men who had lived through the terrible night in the Black Hole. He was a rich man, Lucia's battered tomb proves it.[94]

Kipling here caricatures the dead Lucia and the "pot-bellied East Indiaman," something he had done to living residents of Calcutta elsewhere in his articles. This act of caricature serves to reduce the characters being represented, just as his depiction of the city serves to reduce its stature.[95] Kipling, of course, avoids mentioning other colonial notables buried at the cemetery, such as the orientalist William Jones. This is hardly surprising, for the focus on an unknown character from the city's early colonial history allows him the leeway to reduce Calcutta's past and present residents along with the city itself, in addition to continuing in his project of articulating it as a secondary space. It appears to suggest that the British are in Calcutta owing to the foolhardiness of Company officials in the past. And here again we can note another moment in the larger disavowal of history that marks Kipling's narrative. The centrality of Calcutta to Britain's colonial enterprise in India, and indeed Asia, is simply disowned. Instead, the relentless negative iteration of the city transforms it, in a rhetorical register, into not just a secondary but also a *second-rate* space. Kipling's diatribe against the English East India Company official, furthermore, hints at a sense of embarrassment of the future bard of the British Empire regarding the mercantile provenance of Britain's imperial mission. This is understandable from a writer who, a decade later, would compose "The White Man's Burden," for the lofty notion of the civilizing mission does not sit well with the more mundane idea of economic profit.

Through the use of the images of filth and degradation, and by invoking the sense of smell, Kipling produces Calcutta as a "disgusting" place. Indeed, it is this sense of disgust that allows Kipling to articulate the space of the colonial city as unique and iconic; to articulate the landscape of Calcutta as an inferior 'other' to London. It is this iconicity of the anti-aesthetic landscape of Calcutta that would, through the course of the twentieth century and beyond, be normed into *the* idea of the city, leading to the production of a fetishized image of Calcutta that would mystify the conditions of its production.

A native metropolis

Even as British depictions of Calcutta turned towards the negative in the latter half of the nineteenth century, the local Bengali inhabitants were stepping up their claims to the city. Sumit Sarkar suggests that the nineteenth-century Bengali elite "seldom wrote about Calcutta, and tended to emphasize the negative features of its life when it did."[96] While this is indeed accurate, as we have seen in the case of *Kalikata Kamalalay* earlier, this trend was beginning to change by the 1860s. A text that illustrates this shift quite strikingly is Bholanath Chunder's travelogue, ***The Travels of a Hindoo to Various Parts of Bengal and Upper India,*** one of the earliest travel accounts in English written by an Indian. Published in 1869, the same year that Smith was writing his report condemning Calcutta's civic infrastructure, it provides accounts of eastern and northern India, especially Delhi, based on the author's travels from 1866. In his travel accounts, Chunder not only sets up Delhi, the capital of the pre-colonial Mughal Empire, in contrast to Calcutta but also extols the virtues of the latter as a concrete manifestation of the modern.[97]

Chunder presents his readers with a survey of the Mughal city's built environments under different rulers, before proceeding to describe Shahjahanabad, the city built by the Mughal emperor Shah Jahan. In his description of Delhi, he holds up the pre-colonial city as an explicit and acute foil to Calcutta. The colonial city has the "advantage in general magnificence" for Delhi, he claims, has nothing like Calcutta's splendid squares, and no places for driving and walking such as the Maidan or the Strand.[98] Chunder goes on to invoke that popular and reified trope of depicting Calcutta – the view of the city from the river – to further disparage the Mughal city: the view of Delhi from across the Yamuna is not "half so grand and striking" as the prospect of Calcutta from across the Ganges.[99] In a hyperbolic flourish that puts him at par with Kipling, Chunder comments that "it is much to be doubted whether, in its best days, Delhi had any such tasteful buildings as *our* Fever Hospital, *our* Metcalfe hall, and *our* classical Mint. It is not fair . . . [to compare] a fallen and a rising city."[100] Notice the use of the plural possessive pronoun "our" to refer to the buildings that constitute the built environment of Calcutta's 'White' Town. The native elite no longer seems to feel out of place in the colonial city, and especially its 'White' Town.

Chunder also comments on Shah Jahan's decision to build a new city and move the ruling dispensation there. Ascribing the Mughal monarch's decision to a fit of imperial whimsy, he uses this to compare the "unreasonable" Mughal regime to the "reasonable" nature of British rule in India. In British India, he notes, money has to be made through "honest and life-long labours – and not by looting . . . [The] health of the viceroy cannot be a reason . . . for the building of a new City."[101] And despite the risks to the health of the Governor-Generals, the capital has remained in Calcutta. This, for Chunder, proves the reasonableness of the British:

> three successive Governors-General shortened their lives, one Financier came and was consigned to the grave, and another broke his health and went home to recruit it, still the removal of the metropolis from Calcutta has not taken place considering the immense interests jeopardized . . . [for] such a removal would be worse than an earthquake or an inundation.[102]

Any attempt to move the capital from Calcutta, Chunder notes with an air of quiet confidence, would not only jeopardize the fortunes of the city's house owners but would also devalue the crores of rupees invested in building Fort William, the Government House, the Town Hall, the High Court, the Bank of Bengal, the Post Office, and "the innumerable palace like buildings of our city would not then retain any value in the estimation of men . . . [and the] The greatest house owner who is now esteemed a *millionaire* would find himself reduced to a provincial gentleman."[103] Calcutta is once again designated as "our city," that is, belonging to the propertied sections of the Bengali elite. And readers are also told that shifting the capital from Calcutta would reduce the homeowners to "provincial" folk. The metropolitan confidence and the disparaging of the provincial in this statement are unmistakable. Chunder's fears about shifting the capital of course appear profoundly prescient – and deeply ironic – in light of the British decision to move the capital of British India from Calcutta to Delhi almost half a century later, in 1911. But, more crucially, his comments display the native elites' comfort with the property regimes and the larger processes of colonial capitalist modernity that instituted them. Chunder's comments also ring with the confidence and pride of the native propertied class, speaking from their metropolitan location having tasted the fruits of colonial capitalist modernity. And what is more, these statements are marked by this class' implicit claims on the urban space of Calcutta. And by the beginning of the twentieth century it would be commonsensical enough to claim, as did Binay Krishna Deb, one of the first native historians of the city, that Calcutta was "as much a European and a Marwari as a Bengalee [sic] town."[104]

The native claims to the space of Calcutta that Chunder's comments inaugurate would, as the nineteenth century gave way to the twentieth, continue to grow louder even as they moved beyond the elite domain. Documents authored by the colonial state show a keen awareness of the city's public asserting themselves, not least in Calcutta's public spaces. Their demanding, if disruptive, presence haunts, for instance, the pages of the annual reports by the curator of Calcutta's Public Gardens. The report from 1910 to 1911 speaks of the demands of landscaping Calcutta owing to growing pedestrian traffic and the onslaught of cows:

> In Dalhousie Square Garden the . . . footpaths . . . proved inadequate to accommodate the traffic passing along them. . . . The general appearance of this Garden is gradually improving. In the Curzon Gardens the lavish display of beds of flowering annuals called for a great amount of labour that however yielded very successful results. The shrubs in the temp beds . . . suffered from the depredations of cattle; and as the surroundings, which includes a rubbish tip, are as unaesthetic as ever, the general effect is not as satisfactory as it might be.[105]

The report from some years later continue to report on the menace of pedestrian traffic on Calcutta's gardens:

> The . . . circumstances . . . are not of the most favorable, for pedestrian traffic pays no particular regard to the paths provided, and with regard to the Eden Garden in particular the section of the public that crowds through the Garden

> when football matches are being played in the neighbourhood shows scant respect for the work of the horticulturist.[106]

The same report also notes that the "Curzon Garden . . . suffered a great deal of damage at the feet of the enormous concourse that assembled on the occasion of Their Imperial Majesties' visit to Calcutta."[107] Some years later the situation appears a touch more cataclysmic, with the report noting that during "the Armistice Celebrations most of the annuals in the Curzon and Dalhousie Square gardens were destroyed by the feet of the rejoicing crowds, and both gardens had to be replanted."[108] The remarks in the reports, and Chunder's comments earlier, underline an emergent way of seeing, experiencing, and being in the colonial city for the native elites. But even as these instances shed light on native contestation of, and claims on, Calcutta's urban space they do not really give us a sense of being at home in the city. For such an emic perspective on Calcutta, we need to turn to Rabindranath Tagore's memoir *Chelebela* [ছেলেবেলা], translated into English as *My Boyhood Days*.

Chelebela recounts, as the English title of the translation suggests, Tagore's boyhood days in Calcutta. Published in 1940, a year before his death, the memoir not only presents readers with glimpses of the author's younger self but also takes them back to the city of the 1860s and 1870s. The Calcutta of the memoir is at once a remembered space and a space of remembrance. Tagore draws out a city that has passed from view, and one that can only be recollected. This sense is conveyed in the opening lines of the text itself when he writes: "The Calcutta where I was born was an altogether old-world place."[109] And the lines that follow immediately after lay out the urban life of the city of his childhood and underline the sense of change:

> There were no trams then, no buses, no motors. Business was not the breathless rush that it is now, and the days went by in leisurely fashion. Clerks could take a good pull at the hookah before starting for office, and chew their betel as they went along. Some rode palanquins, others joined in groups of four or five to hire a carriage in common. . . . Wealthy men had . . . their [own] carriages.[110]

The passage above tells us about changes in the city's transport systems that have, in turn, fundamentally transformed urban life, especially its temporal rhythms. The city that emerges from this description is one that is slower, quieter, and less in the thrall of the modern than its present incarnation. Indeed, throughout the memoir, Tagore contrasts the remembered, late-nineteenth-century Calcutta to the city of the present stressing the changes to urban life, which becomes a recurring trope in his memoir.

Yet Tagore does more than simply narrate the alteration of urban life; he portrays the change as it was lived by Calcutta's residents. To put this another way, Tagore's memoir shows the transformation in the ways of being in the urban space. Take, for instance, his poignant comments on the arrival of artificial light in Calcutta. He narrates this not only, or simply, as the conquest of darkness by the forces of colonial capitalist modernity.[111] Instead, we see him situating the arrival of light to changes in the way life is lived in the city. He writes that there was no gas or electric light in the city of his childhood, and when the kerosene lamp was introduced, "its brilliance amazed us."[112] These changes have concrete effects on lives both within and outside the home. At home, the advent of artificial light meant studying longer hours and he reminisces about being taught from "Peary Sarkar's First Book" in that light.[113] And beyond the home, it meant for a city that was less sleepy:

> When I was a little boy Calcutta city was not so wakeful at night as it is now. Nowadays, as soon as the day of sunlight is over, the day of electric light begins. . . . The oil mills are still, the steamer sirens are silent, the labourers have left the factories, the buffaloes which pull the carts of jute bales are stabled in the tin roofed sheds. But the nerves of the city are throbbing still with the fever of thought which has burned all day in her brain. Buying and selling go on as by day in the shops that line the streets. . . . Motors continue to run in all directions, emitting all kinds of raucous grunts and groans, though they no longer run with the zest of the morning. But in those old times which we knew, when the day was over whatever business remained undone wrapped itself up in the black blanket of the night and went to sleep in the darkened ground-floor premises of the city.[114]

This description of Calcutta straddles the present and the past, allowing Tagore to showcase changes in the daily life of the city and its people. Focusing on the city's worlds of work, and workers, he shows us how the advent of artificial light meant the lengthening of the working day and how this, in turn, changed the way locals lived and inhabited the city leading to a transformation in Calcutta's urban rhythms. Notice, also, that we are presented with *views* of the city bustling with business after dark; of the *sirens* of steamers and of the *grunts* and *groans* of motorcars; and we read about the *zest* of the city, its *throbbing* nerves, and the day's undone business being *wrapped* up in the black blanket of night. Readers, then, are presented with visual, aural, and tactile registers of the city: Calcutta is seen, heard, and felt. In sum, this is an articulation of the city as it is lived, and experienced through the senses, as a totality. I draw attention to this as it shows Tagore delineating an immanent and situated perspective on the city. This is akin to Kipling's narratorial position of *witnessing*, but at once singularly different from it. Tagore does not simply witness the city but articulates a lived relationship to it. And this is also what makes his perspective strikingly distinct from Bholanath Chunder's. For Tagore, unlike Chunder, the city is not an agglomeration of buildings or a site of investment. Neither does he champion it as a space of modernity, holding it up as distinct from other spaces. Instead, it is the site of childhood and growing up for him. Simply put, it is home, and Tagore's writings underscore the native elites' sense of being at home in it. Recall Bhabanicharan Bandyopadhyay's unease with the city from a century ago. Tagore's descriptions show a radical transformation in the native elites' attitudes towards the colonial city. In fact, they gesture towards changes in the ways of being in the city.

Tagore's narration of Calcutta, furthermore, presents images of the city as a space that cuts across social groups and classes. Readers are presented with sketches of the sweeper Sobha Ram, "a wrestler, [who also] . . . spend[s] a good deal of time in practising his preparatory feints and approaches, and in brandishing his heavy clubs."[115] We also hear of "Abdul the sailor, with his pointed beard, shaven moustache and close-cropped head . . . [who] brings hilsa fish and turtle eggs from the Padma."[116] Describing the people in and around his household, Tagore writes:

> I could hear the shouts of the various servants of their work. Pari the maid is returning from the bazaar through the front courtyard with her vegetables in a basket on her hip. Dukhon the bearer is carrying in Ganges water in a yoke across his shoulder. The weaver woman has gone into the inner apartments to trade the newest style of saris. Dinu the goldsmith, who receives a monthly wage, usually sits in the room next to the lane, blowing his bellows and carrying out the orders of the family; now he is coming to the counting house to present his bill to Kailash Mukherjee, who has a quill pen stuck over his ear. The carder sits in the courtyard cleaning the cotton mattress. . . . Mukundalal the durwan [doorman] is rolling on the ground outside with the one-eyed wrestler, trying out a new wrestling fall . . . he slaps his thigh loudly. . . . There is a crowd of beggars sitting waiting for the regular dole.[117]

This sketch shows the young child looking in on the workers, and the goings-on of his aristocratic household. The passage above registers, in particular, the multi-ethnic character of the space of the household and, by extension, the urban space. Tagore, in fact, details the people as they go about doing their work with an ethnographic eye that is once again reminiscent of Kipling. While Tagore shares a certain ethnographic modality with Kipling, his depictions of the residents are nevertheless profoundly different. Unlike Kipling, Tagore does not sketch out Calcutta as a museum filled with native types. Neither are his depictions of the people in his memoir structured by a desire, as in Kipling, to classify and categorize. Instead, they are delineated as real and living people, with these individuals and groups *interacting* with each other. Sobha Ram the sweeper is shown being teased by boys who shout the name of Hindu gods to his ears. Abdul tells fantastical stories, while Dinu the goldsmith is sketched as the person who "receives a monthly wage," sits in a room and blows "his bellows . . . carrying out the orders of the family." Likewise, Kailash Mukherjee, who is shown interacting with Dinu, is spoken of as a man with a "quill pen stuck over his ear." And Mukandaram, the doorman, is shown slapping his thighs even as he practices his wrestling moves. Tagore's descriptions are suffused with a distinct sense of empathy that humanizes the people he represents. They are shown as individuals with quirks who go about their lives to meet the needs and demands of their daily existence. In so doing, Tagore presents Calcutta as a lived, and living, space and this provides a distinctiveness to his depictions.

Tagore's writings from the mid-twentieth century bring to a close my consideration of the various depictions of the city from the early decades of the nineteenth century onwards. As we have seen, British responses to the city ranged from adulation to contempt and outright disgust. Indians, on the other hand, initially appear to have been bewildered by the urban space that bespoke spatialized alienation. From the end of the nineteenth century, however, Indians, especially the natives of Bengal, saw beyond Calcutta's colonial provenance to claim it as their own. In sum, these imagined urban geographies range from presenting Calcutta as a space of wonder and beauty to one of exceptional filth and squalor; from a "heartless" and bewildering city, to a space of modernity and home. These distinct, and at times fundamentally opposed, ways of perceiving, conceiving, and being in, the city signal diverse expressions of spatial desire that undergird the contestation of the city's space at distinct moments of its history. But this story of *spatial struggle* and desire is also, and at once, a story of a desire for *scalar politics*. The *local* story of the claims and counter-claims over the space of Calcutta, the centre of Britain's Indian empire, also gestures towards the larger struggles over the space of the colony, that is, towards decolonization.

Notes

1 See Kapil Raj, "The Historical Anatomy of a Contact Zone: Calcutta in the Eighteenth Century," *The Indian Economic and Social History Review* 48, no.1 (2011): 154–179.

2 B. K. Deb, *The Early History and Growth of Calcutta* (Calcutta: Romesh Chandra Ghose, 1905), 1.

3 Henri Lefebvre, *Writings on Cities*, trans. Eleonore Kofman and Elizabeth Lebas (Malden: Blackwell Publishing, 1996), 102.

4 P. T. Nair, "The Growth and Development of Old Calcutta," in *Calcutta: The Living City, Volume I: The Past*, ed. S. Chaudhuri (Calcutta: Oxford University Press, 1990), 11. See also, P. T. Nair, *Calcutta in the 17th Century* (Calcutta: Firma KLM, 1986).

5 The Nawab's officers supposedly confined British residents in a small room within Fort William, some of who died from suffocation the next day. Though this was never independently confirmed, it was exaggerated into the "Black Hole" of Calcutta and invoked through the eighteenth to twentieth centuries as evidence of Indian brutality against the British. This would also mould British perceptions about Calcutta negatively in the years to come. See K. Telscher, "India/Calcutta: City of Palaces and Dreadful Night," in *The Cambridge Companion to Travel Writing*, eds. P. Hulme and T. Youngs (Cambridge: Cambridge University Press, 2002), 191–206. For a detailed study of how the supposed Black Hole tragedy paved the ideological foundations for British imperial rule in India and its "civilizing" mission, see Partha Chatterjee, *The Black Hole of Empire: History of a Global Practice of Power* (Princeton, NJ: Princeton University Press, 2012).

6 Nair, "Growth and Development," 12–13.

7 Henry Beverly, *Report on the Census of the Town of Calcutta Taken on the 6th April 1876* (Calcutta: Bengal Secretariat Press, 1876), 40.

8 Pradip Sinha, *Calcutta in Urban History* (Calcutta: Firma KLM, 1978).

9 Sumanta Banerjee, *The Parlour and the Streets: Elite and Popular Culture in Nineteenth Century Calcutta* (Calcutta: Seagull Books, 1989), 29.

10 Scholars of colonial Calcutta have repeatedly demonstrated the contradictory nature of the city's landscape as evident in the development of the city's 'White' and 'Black' Towns. See Sumit Sarkar, *Writing Social History* (New Delhi: Oxford University

Press, 1997), 159–185; S. N. Mukherjee, *Calcutta: Essays in Urban History* (Calcutta: Subarnarekha, 1993); Sinha, *Calcutta in Urban History*; and Anthony D. King, *Colonial Urban Development: Culture, Social Power and Environment* (London: Routledge and Kegan Paul, 1976). This model has been critiqued by Swati Chattopadhyay who argues against, and indeed rejects, the notion of the 'White' and 'Black' Towns. See Swati Chattopadhyay, *Representing Calcutta: Modernity, Nationalism, and the Colonial Uncanny* (London: Routledge, 2005) and Swati Chattopadhyay, "Blurring Boundaries: The Limits of 'White Town' in Colonial Calcutta," *Journal of the Society of Architectural Historians*, 59 (2000): 154–179. Though a substantial engagement with her claims is beyond the scope of this chapter, it bears mentioning that her intervention has spawned an entire body of scholarship celebrating spatial and racial fluidity, indeterminacy, and ambivalence. For instance, Preeti Chopra's study on colonial Bombay takes up Chattopadhyay's notion of fluidity to argue that "British Bombay was envisioned and built jointly by colonial rulers and Indian and European mercantile elites to serve their various interests" (xv). See Preeti Chopra, *A Joint Enterprise: Indian Elites and the Making of British Bombay* (Minneapolis, MN: University of Minnesota Press, 2011).

11 The old fort was repaired and used as a customs house from 1766 onwards.

12 H. J. Rainey, *A Historical and Topographical Sketch of Calcutta* (Calcutta: "Englishman" Press, 1876), 61.

13 The new Fort William was shaped as a rough octagon. Five sides of the fort faced the Maidan, and the other three the river. See Swati Chattopadhyay, "Cities and Peripheries," *Historical Research* 83, no. 222 (2010): 649–671.

14 Chattopadhyay, "Blurring Boundaries," 167.

15 G. N. Curzon, *British Government in India: The Story of the Viceroys and Government Houses, Volume I* (London: Cassell, 1925), 28.

16 See Banerjee, *The Parlour and the Streets*, and Sarkar, *Writing Social History*.

17 Sarkar, *Writing Social History*, 166.

18 Ibid.

19 H. Chakrabarti, "The Victoria Memorial," in *Calcutta: The Living City, Volume I: The Past*, ed. S. Chaudhuri (Calcutta: Oxford University Press, 1990), 256.

20 Telscher, "India/Calcutta," 195.

21 James Atkinson, *The City of Palaces: A Fragment, and Other Poems* (Calcutta: Government Gazette Press, 1824), 5.

22 See, for instance, Johan Stavorinus, *Voyages to the East Indies, Vol. I* (London: G. G. & J. Robinson, 1798), 122–123; William Hickey, *Memoirs of William Hickey, Vol. 2*, ed. Alfred Spencer (London: Hurst & Blackett, 1916), 120; Fanny Parks, *Wanderings of a Pilgrim in Search of the Picturesque, Vol. II*, ed. Esther Chawner (Karachi: Oxford University Press, 1975), 102; Eliza Fay, *Original Letters from India (1779–1815)*, ed. E. M. Forster (London: Chatto & Windus, 1986), 171–172.

23 Atkinson, *The City of Palaces*, 7. The word "ghat" (spelt "Ghaut" here) refers to the stepped landings for boats on the banks of inland water bodies, especially rivers.

24 Ibid., 15.

25 Ibid.

26 Ibid., 28.

27 Ibid., 12–21.

28 The 1850 census of Calcutta notes the types of buildings of the city. It records 13,120 "Pucka Houses" – that is, built of bricks or stone – and 49,445 "Native Huts" in the city. See F. W. Simms, *Report of the Survey of Calcutta* (Calcutta: Military Orphan Press, 1851), 41–43.

29 Quoted in Ralph Russell and Khurshidul Islam, *Ghalib: Life and Letters* (New Delhi: Oxford University Press, 1994), 48.

30 For a study that situates the *shahrashub* in relation to European literary traditions, see Michele Bernardini, "The *masnavi-shahrashub* as Town Panegyrics: An International

Genre in Islamic Mashriq," in *Narrated Space in the Literature of the Islamic World*, eds. Roxanne Haag-Higuchi and Christian Szyska (Wiesbaden: Harrassowitz, 2001), 81–94.

31 Sunil Sharma, "The City of Beauties in Indo-Persian Poetic Landscape," *Comparative Studies of South Asia, Africa and the Middle East* 24, no. 2 (2004): 73–74.

32 The *masnavi* is an Indo-Persian genre of indefinite length with an aa-bb rhyme scheme and is usually classified as didactic or a romance. Prosodically indigenous to Persia, its most famous exemplar is Firdausi's *Shahnama* from the eleventh century. For more on the *masnavi* see Julie Scott Meisami, *Medieval Persian Court Poetry* (Princeton, NJ: Princeton University Press, 1987). For a study on the late classical use of the romance *masnavi*, see Prashant Keshavmurthy, *Persian Authorship and Canonicity in Late Mughal Delhi: Building an Ark* (London: Routledge, 2016).

33 Sharma, "The City of Beauties," 76.

34 Ibid., 77.

35 Mirza Ghalib, "Conversation with his Guiding Angel," in *The Shadow of a Bird in Flight*, ed. and trans. Shamsur Rahman Faruqi (New Delhi: Rupa, 1994), 45.

36 Mirza Ghalib, *Divan-i Ghalib Dihlavi*, ed. Muhsin Kiyani (Tehran: Rawzanah, 1997), 389. I am grateful to Sunil Sharma for helping me locate this text and to Prashant Keshavmurthy for translating it.

37 The title *Kalikata Kamalalay* [কলিকাতা কমলালয়] literally translates into "Calcutta, the Abode of Kamala." Kamala is another name for Lakshmi, the Hindu goddess of wealth and the tract, in its title, draws attention to the city's economic status and power.

38 Bhabanicharan Bandyopadhyay, *Kalikata Kamalalay*, trans. A. K. Roy (Calcutta: Firma KLM, 1990), xi.

39 Ibid.

40 Ibid., xiii–xiv.

41 Edwin Hodder, *Cities of the World: Their Origin, Progress, and Present Aspect* (New York, NY: Cassell & Co., 1886), 62.

42 Ibid., 175.

43 *First Annual Report of the Sanitary Commission for Bengal, 1864–65* (London: Unknown, 1866), 61.

44 David Smith, *Report on the Drainage and Conservancy of Calcutta* (Calcutta: Bengal Secretariat Press, 1869), 4, italics in original.

45 P. T. Nair, "Civic and Public Services in Old Calcutta," in *Calcutta: The Living City, Volume I: The Past*, ed. S. Chaudhuri (Calcutta: Oxford University Press, 1990), 229.

46 Ibid., 227.

47 Ibid.

48 Hodder, *Cities of the World*, 171.

49 Ibid.

50 Ibid., 175.

51 The most notable of these texts are perhaps Henry Mayhew's *London Labour and London Poor*, serialized initially during the 1840s in the *Morning Chronicle* and subsequently published in 1851 in three volumes; and Friedrich Engels's *The Condition of the Working Class in England* published in 1845. See Henry Mayhew, *London Labour and the London Poor, Volumes 1–4* (New York: Dover Publications, 1968); and Friedrich Engels, *The Condition of the Working Class in England: From Personal Observations and Authentic Sources* (Moscow: Progress Publishers, 1973).

52 Andrew Mearns, *The Bitter Cry of Outcast London* (London: James Clarke & Co., 1883), 4.

53 A classic study of the lower classes, and inter-class relationships, of Victorian London remains Gareth S. Jones, *Outcast London: A Study in the Relationship Between Classes in Victorian Society* (Oxford: Clarendon Press, 1971). For the classed and gendered nature of sanitary and sexual regulations, as well as Victorian thinking on dirt and filth, see Seth Koven, *Slumming: Sexual and Social Politics in Victorian*

London (Princeton, NJ: Princeton University Press, 2004); Anne McClintock, *Imperial Leather: Race, Gender, and Sexuality in the Colonial Contest* (New York: Routledge, 1995); and Judith R. Walkowitz, *Prostitution and Victorian Society: Women, Class, and the State* (Cambridge: Cambridge University Press, 1980).

54 E. Mitchell, *Thacker's Guidebook to Calcutta: Its High-Ways & By-Paths, With a Chapter on the Government of India* (Calcutta: Thacker, Spink & Co., 1890) 1–112.
55 Ibid., 113–224.
56 Ibid., 196.
57 Ibid., 199.
58 Ibid., 201–202.
59 Rudyard Kipling, *The City of Dreadful Night and Other Places* (London: Sampson Low, Marston & Co., 1899), 7.
60 Ibid.
61 Ibid., 7–8.
62 Ibid., 8.
63 Ibid.
64 The nineteenth century witnessed the rapid industrialization of the Calcutta hinterland, particularly with the development of the jute industry. For details see Sugata Bose, *Peasant Labour and Colonial Capital: Rural Bengal Since 1770* (Cambridge: Cambridge University Press, 1993). Also see Subho Basu, *Does Class Matter?: Colonial Capital and Workers' Resistance in Bengal, 1890–1937* (New Delhi: Oxford University Press, 2004). For an ethnographic study of the Howrah and the surrounding post-industrial spaces, see Atreyee Majumder, *Time, Space and Capital in India: Longing and Belonging in an Urban-Industrial Hinterland* (London: Routledge, 2018).
65 Kipling, *The City*, 9.
66 Ibid.
67 Kate Flint, *The Victorians and the Visual Imagination* (Cambridge: Cambridge University Press, 2000), 2. Also see Chris Otter, *The Victorian Eye: A Political History of Light and Vision in Britain, 1800–1910* (Chicago: University of Chicago Press, 2008).
68 Christopher Pinney, *Camera Indica: The Social Life of Indian Photographs* (London: Reaktion Books, 1997), 17–71.
69 See Zahid Chaudhury, *Afterimage of Empire: Photography in Nineteenth-Century India* (Minneapolis, MN: University of Minnesota Press, 2012).
70 Johannes Fabian, *Time and the Other: How Anthropology Makes its Object* (New York: Columbia University Press, 1983), 1–36.
71 Kipling, *The City*, 10.
72 Ibid., 10–11.
73 Ibid., 11.
74 A possible exception to this is arguably his depiction of Hurree Mookerjee of *Kim*.
75 Kipling, *The City*, 29.
76 Ibid., 43–44.
77 Ibid., 44.
78 Ibid., 57.
79 Ibid., 63.
80 Ibid., 59.
81 Ibid.
82 Ibid., 61.
83 Ibid., 60–61.
84 Ibid., 72.
85 Ibid., 73.
86 Ibid.
87 Ibid.
88 Ibid.

89 See Peter Stallybrass and Allon White, *The Politics and Poetics of Transgression* (Ithaca: Cornell University Press, 1986), 125–148.
90 Kipling, *The City*, 82.
91 See Thomas Metcalf, *An Imperial Vision: Indian Architecture and Britain's Raj* (Oxford: Oxford University Press, 2002). Also see Narayani Gupta, *Delhi Between Two Empires, 1803–1931: Society, Government and Urban Growth* (New Delhi: Oxford University Press, 1981).
92 Kipling, *The City*, 83–84.
93 Ibid., 86.
94 Ibid., 87–88.
95 See Ernst H. Gombrich, "Cartoonists' Armoury," in *Meditations on a Hobby Horse: And Other Essays on the Theory of Art* (London: Phaidon, 1971), 127–142.
96 Sarkar, *Writing Social History*, 177.
97 For an extensive examination of the place of Delhi in the writings of the Bengali bhadralok in the late-nineteenth and twentieth centuries, see Sandeep Banerjee and Subho Basu. "The City as Nation: Delhi as the Indian Nation in Bengali bhadralok Travelogues 1866–1910," in *Cities in South Asia*, eds. Crispin Bates and Minoru Mio (Abingdon: Routledge, 2015), 125–142.
98 Bholanath Chunder, *The Travels of a Hindoo to Various Parts of Bengal and Upper India, Volume II* (London: N. Trübner, 1869), 375.
99 Ibid.
100 Ibid., 376, italics added.
101 Ibid., 273.
102 Ibid., 272–273.
103 Ibid., 272, italics in original. A "crore" is a unit in the South Asian numbering system that is equal to ten million.
104 Deb, *The Early History*, 3.
105 Bengal Secretariat, *Report on the Royal Botanical Gardens and other gardens in Calcutta, and the Lloyd Botanic Garden in Darjeeling for the year 1907–08* (Calcutta: Bengal Secretariat Press, 1908), 3–4.
106 Bengal Secretariat, *Report on the Royal Botanical Gardens and other gardens in Calcutta, and the Lloyd Botanic Garden in Darjeeling for the year 1911–12* (Calcutta: Bengal Secretariat Press, 1912), 3.
107 Ibid.
108 Bengal Secretariat, *Report on the Royal Botanical Gardens and other gardens in Calcutta, and the Lloyd Botanic Garden in Darjeeling for the year 1918–19* (Calcutta: Bengal Secretariat Press, 1919), 2.
109 Rabindranath Tagore, *My Boyhood Days*, trans. Marjorie Sykes (Calcutta: Viswabharati, 1945), 3.
110 Ibid.
111 See Wolfgang Schievelbusch, *Disenchanted Night: The Industrialization of Light in the Nineteenth Century* (Berkeley: University of California Press, 1995).
112 Tagore, *My Boyhood Days*, 5.
113 Ibid. By "Peary Sarkar's First Book," Tagore here refers to Peary Charan Sircar's *First Book of Reading for Native Children*, the first in a series of six introductory texts on the English language for Indians. It was first published in 1850.
114 Ibid., 21.
115 Ibid., 4.
116 Ibid., 10.
117 Ibid., 9.

3 That magnificent song

Between the performative and the pedagogic

On 24 January 1950, Rajendra Prasad, presiding over the Constituent Assembly of India, issued a statement on the adoption of the national anthem for the newly independent nation-state. He noted that the

> composition consisting of words and music known as Jana Gana Mana [sic] is the National Anthem of India, subject to such alterations as the Government may authorize as occasion arises, and the song Vande Mataram [sic], which has played a historic part in the struggle for Indian freedom, shall be honored equally with Jana Gana Mana [sic] and shall have equal status with it.[1]

Prasad's comment marks the incorporation of these two patriotic lyrics – or, more accurately, their truncated versions – as state symbols of the Indian republic, setting off the postcolonial careers of Rabindranath Tagore's *Jana Gana Mana* [জনগণমন] (first published in 1911) and Bankim Chandra Chatterji's *Vandemataram* [বন্দেমাতরম] (first published as a poem in the journal *Bongodorshon*, in 1875; subsequently published as a part of *Anandamath*, in 1882) as the country's national anthem and national song respectively.[2] They also mark the culmination of a debate about the appositeness of *Vandemataram* as a state symbol of secular India. This was because Chatterji's *Vandemataram* gestures towards an ethno-religious nationalism that draws on the Hindu iconographic tradition to privilege a Hindu identity for the postcolony.[3] By contrast, Tagore's *Jana Gana Mana* posits South Asia as an inclusive civilizational space for all ethnic and religious communities. Despite the parochialism of its vision, *Vandemataram* did indeed play, as Prasad notes, a "historic part" in the Indian National Congress–led (hereafter, INC) anti-colonial struggle. The lyric presents the Bengal landscape as a lived space by rendering it as an incarnation of the Hindu mother goddess. As the song gained popularity across the British Indian colony, its spatial reference came to signify India instead of Bengal. This was owing to the song's phenomenal popularity across British India after the Congress adopted the practice of singing it at the start of its meetings. Despite being proscribed by the British at specific times, the song maintained a firm hold on the popular anti-colonial imagination with the phrase "vandemataram," which translates to "I revere the mother," becoming a stand-alone phrase used as an anti-colonial slogan and greeting.[4] The song, as several scholars have noted, also aided the development of the anthropomorphic imagination of the British Indian colony as a specific – and gendered – geo-body, that is, Bharat Mata or Mother India.[5]

Vandemataram, however, was not without its share of controversies. Indian Muslims, since the early decades of the twentieth century, began criticizing the lyric for articulating a Hindu majoritarian view of India. This perspective on the lyric, which hardened through the years, appears to be supported by the overtly anti-Muslim rhetoric of the novel *Anandamath* of which *Vandemataram* is a part.[6] In the Calcutta riots of 1921, "vandemataram" emerged as a slogan for Hindu rioters against Muslims, eventually becoming, as one historian notes, a sort of war-cry for Hindu fanatics.[7] Two and a half decades later, in 1937, the All India Muslim League marked its antipathy against this "highly objectionable song" in a resolution, calling the INC's attempts to introduce it as the national anthem of the country "callous, positively anti-Islamic, [and] idolatrous in its inspiration and ideas."[8] But *Vandemataram* had other critics too. The most prominent among them was Tagore, colonial India's pre-eminent literary figure and public intellectual. Ironically enough, Tagore had introduced *Vandemataram* to the INC by setting it to music and singing it – the first instance of its public performance – at the INC's 1896 session in Calcutta.[9] But in the aftermath of the Hindu-Muslim riots that followed the 1905 Partition of Bengal, Tagore increasingly became a vocal critic of the deified imagination of the nation.[10] In 1937, Subhash Chandra Bose, then the President of the INC, wrote to Tagore asking for his opinion on the appositeness of *Vandemataram* as the national anthem of India. The poet's response on the question was emphatic:

> The core of Vande Mataram [sic] is a hymn to goddess Durga: this is so plain that there can be no debate about it. Of course Bankimchandra does show Durga to be inseparably united with Bengal in the end, but no Mussulman [Muslim] can be expected patriotically to worship the ten-handed deity as 'Swadesh' [our native land]. . . . The novel *Anandamath* is a work of literature, and so the song is appropriate in it. But Parliament is a place of union for all religious groups, and there the song cannot be appropriate.[11]

The Constituent Assembly's decision to select Tagore's *Jana Gana Mana* as independent India's national anthem was an awkward resolution of the question of *Vandemataram*'s place in the life of the postcolonial nation-state. While the decision was undoubtedly a conciliatory gesture to India's Muslims, the Constituent Assembly's enshrining of *Vandemataram* as the National Song shows them acknowledging the lyric as an iconic representation of Indian space.

I lay out the history of the socio-political reception of *Vandemataram* here not only because it is an important interpretive context for the lyric. More significantly, it illuminates the normative trend in critical scholarship on the lyric, by no means insubstantial, that has focused almost exclusively on reading it as a historical document, and in terms of its particular effects on the Indian anti-colonial struggle.[12] These insights undoubtedly illuminate the worldliness – understood in Edward Said's sense of the term – of *Vandemataram* that makes the lyric legible in terms of the social history of Hindu-Muslim relations and Hindu nationalism in colonial (and postcolonial) India.[13] Nevertheless, by situating the lyric in determinate local contexts of history and politics, these reading practices set out overdetermined ways of interpreting it. And in so doing, they remain incurious of another register of worldliness: the lyric's mediation, in the aesthetic domain, of a global politics of space-making structured by colonialism and anti-colonialism.

In what follows, I read *Vandemataram* – and *Jana Gana Mana* as well – in a somewhat unfamiliar and defamiliarizing way seeking to understand their spatial desires, and how they articulate their counter-colonial spatial imaginaries. I constellate these two lyrics for a number of reasons, not least because both their truncated versions enjoy the prestige of state symbols in postcolonial India. But more crucially for my purposes, the lyrics share some marked similarities in content and form. Both re-imagine the colony as a postcolony and are thus suffused with counter-colonial spatial desire. And while both structure their imaginations of space as landscapes, they also envision the utopian space of the postcolony, and ways of belonging to it, in fundamentally different ways. As will become evident, *Vandemataram* idealizes the famine-afflicted Bengal landscape as a space of fecundity and draws on popular Hindu iconographical traditions to idolize it as a mother goddess. Its vision of spatial belonging is fashioned through Hindu conceptual protocols and posits Indians in a relationship of *filiation* with that space, and each other. Tagore's *Jana Gana Mana*, on the other hand, posits a clear distinction between divinity and India, gesturing instead to the Indian landscape as an inclusive civilizational space for all ethnic and religious communities. Tagore's India is space of *affiliation*, where Indians are bound together by a common habitus structured by a shared history and geography.[14]

Both lyrics invoke the "spiritual" to structure their visions of the postcolony but, once again, in distinct ways. Readers will recall that Partha Chatterjee locates the emergence of Indian nationalism – and, in fact, anti-colonial nationalisms in Asia and Africa – within an indigenous "spiritual" realm that he suggests is in contradistinction to the colonial "material" domain.[15] These songs show that the "spiritual" was indeed a crucial resource for imagining the postcolony. However, the lyrics' differing and divergent ways of deploying the "spiritual" demonstrates the plurality of that category. Pluralizing that category, then, necessarily provides the category of "difference," so central to Chatterjee's conceptualization of anti-colonial nationalism, with content and specificity and rescues it from turning into its opposite, that is, the "monotony of the Same."[16] Moreover, the "spiritual" is in no way unique to, or even especially true of, imagining the anti-colonial or peripheral nation – or, in my telling, the postcolony. As will become clear from my readings of English patriotic lyrics, it also structures articulations of metropolitan spaces and nationalisms. Furthermore, both *Vandemataram* and *Jana Gana Mana* belong to a global genre that I shall be calling the *patriotic lyric*. *Vandemataram* is one of the earliest patriotic lyrics from South Asia that, through translation and imitation into other South Asian languages, served as a catalyst for popularizing the genre across British India. As such, *Vandemataram* and *Jana Gana Mana* exemplify South Asian instantiations of the global form of the patriotic lyric, which I will delineate before proceeding with my discussion of the two lyrics.

Lineaments of a global genre

Literary studies scholarship typically understands the novel as a global and modern genre. Yet, alongside, and analogous, to the rise of the novel, it is also possible to chart the development of another form that can be understood as a lyrical response to the uneven processes of colonial capitalist modernity. This is the patriotic lyric that emerges coevally with the novel and, over time, becomes a recognizable genre across the globe. Recall that we live in a world-system of nominally equal nation-states, each with its own national anthem. The specific form that I am outlining here would include all these anthems, but also a larger body of lyrics of which they are a part and from which they are drawn. Let me illustrate my point with a few specific examples. In the case of Britain, this genre would include "God Save the Queen/King" of course, but also lyrics such as James Thomson's "Rule! Britannia" as well as William Blake's "Jerusalem," so beloved of English football fans. In the Indian context, this corpus would comprise Tagore's *Jana Gana Mana* and Chatterji's *Vandemataram*, both lyrics in the Bengali language. Other Indian instantiations of the genre would have to include Muhammad Iqbal's *Taraana-i-Hindi*, the iconic Urdu song popular to Indians through its opening line: "saare jahan se accha, Hindustan hamaara" ("Better than the whole world, is our India") as well as that Hindi film song that is one of the most effusive celebrations of post-independence India that goes "merey desh ki dharti sona ugley, ugley hirey moti" [मेरे देश की धरती सोना उगले उगले हीरे मोती] and translates as "my country's soil brings forth gold, it brings forth diamonds and pearls."[17]

These songs – and this is true of similar songs from other geographical contexts as well – all share a set of defining characteristics of which the most significant is the depiction and valorization of specific spaces and people. The songs imagine and articulate conceptions of space and spatial belonging that, through the sedimentation of this socio-spatial content, engenders a distinct form. A few other characteristics of this genre must also be adduced. These patriotic lyrics are "lyrics" in the originary sense of the term. They are typically sung, making them a performative poetic form. Stridently public in tone, these songs are also self-consciously interpellative, situating their auditors in a relationship – often as insiders or outsiders – to the spaces they imagine. In sum, I am suggesting that patriotic lyrics, in their form and function, are invested in a project of socio-spatial performance and pedagogy.[18] They imagine, perform, and teach about spaces, and ways of being in those spaces.

This is of course not to claim that other poetic forms are not marked by the socio-spatial, or that they do not also teach us about spaces and their inhabitants, for they certainly do. But patriotic lyrics I will insist are distinct from all others because they depict, and perform, not local *places*, but *spaces* in the abstract. It is helpful to recall here my discussion on space-making from the Introduction. I have suggested that the global spatial formation and re-formation under conditions of colonial capitalist modernity, exemplified by the development of the singular and uneven world-system and the attendant processes of colonization and decolonization, produce *new* – and *new kinds* of – spaces such as the colony, metropole, and the postcolony. The form of the patriotic lyric, I am suggesting, is a registration of this spatial newness in the aesthetic domain. It is a formal response to the felt social need for an aesthetic education about the new spaces forged by capitalist modernity and ways of inhabiting them. Since the abstract spaces of modernity such as the nation or the colony are unrepresentable as such, these patriotic lyrics employ a range of strategies to re-present them. They make concrete categories stand in for these abstract spaces. Among the most common are representing the spaces of the colony or the nation through, and as, a spatial view or *landscape*; alternatively, these spaces are represented through an *anthropomorphic* form (for instance, Britannia for Britain; Uncle Sam for the United States; Mother India for India, and so on). In both cases, the concrete image of a landscape or a human figure is made to stand in for the abstract spaces thus constructing a relationship of non-identical similitude between the concrete image and the abstract space. In other words, they represent these abstract spaces in an *allegorical* manner.[19]

To re-present the abstract as the concrete is also to re-signify the mediated with a sense of immediacy. Patriotic songs infuse *space* with an experiential register, seeking to re-enchant and aesthetically transform it into a *place* of community, and of communitarian dwelling. Key to understanding this social function of the form is to appreciate their performative nature. As a sung genre, patriotic lyrics approximate the earlier notion of poetry as practice, and indeed, praxis, underlining a way in which orature endures in a world of literature.[20] In their singing, which is often a social act as it is done collectively, the lyrics not only evoke and disseminate an imagination of the space but also affectively situate the singers and auditors as subjects in that landscape. In their interpellative address, the songs formulate and perform a sense of community, and settle questions of belonging. This performance of spatial subjecthood is complemented by actual bodily practice – the acts of singing and playing music – and introduce the experiential not just into the lyric but also, more crucially, into the imagination of space and community it engenders. This performative character, which accords the genre its distinctiveness and ideological power, enables it to be an imperative form that instantiates spatial imaginaries, identities, and subjectivities, as if charming them into being. Patriotic songs therefore recode the artwork with its originary ritualistic element ordaining it with the power to re-enchant the disenchanted spaces and subjects of capitalist modernity.[21]

It is unsurprising that, as a distinctly socio-spatial genre, patriotic songs register the spatial unevenness of capitalist modernity. This is most evident in the way they mediate imperialism, and the attendant logic of spatial difference, in their content and form. In metropolitan locations, they typically take on a colonial charge to develop as what Suvir Kaul has termed "anthems of empire," and mutate over time into songs of metropolitan nations.[22] In peripheral locations, they often – though not always, or even necessarily – take on a critical stance on imperialism.[23] These lyrics re-imagine the peripheral space in an anti-colonial vein, re-presenting the colony as the postcolony. Anti-colonial patriotic songs, then, perform a dual function. They too are, like their metropolitan counterparts, tasked with re-enchanting space but they must also, and at once, refuse colonial imaginaries of peripheral space. This nexus of socio-spatial negation and re-enchantment marks the anti-colonial patriotic lyrics with a spatial desire and makes them key means for imagining the postcolony.

The colonial patriotic lyric

James Thomson's "Rule, Britannia" is surely one of the more resonant exemplars of the colonial patriotic lyric. Since its first performance in 1740, as the finale to *Alfred: A Masque*, Thomson's lyric has orchestrated a "theological, political, cultural, and even civilizational authority [that articulates and even] enacts fully the desire of British poets for a cultural power that would be more than literary."[24] And it has since, while urging Britain to "rule the waves," proclaimed that "Britons never will be slaves."[25] The lyric's glorification of Britain and Britons is in many ways typical of the colonial patriotic lyric. And it is precisely its typicality that makes it remarkable, for it provides us with the colonial lineaments of the genre marked by a distinct imagination of space, and global spatial domination. The poem opens with the lines:

> When Britain first, at Heaven's command
> Arose from out the azure main;
> This was the charter of the land,
> And guardian angels sang this strain:
> "Rule, Britannia! rule the waves:
> Britons never will be slaves."[26]

The lines, as Kaul notes in his reading of the poem, present Britain's origins as divinely ordained while at the same time yoking the spiritual with the more mundane domains of the legal and commercial.[27] Moreover, the image of Britain emerging "from out the azure main" signals a specific spatial perspective. It presupposes a transcendental, that is, a separated and distanced way of seeing and showing, one that enables the perception of Britain, in one glance, as a spatial unit and a unitary space. These protocols of spatial representation and visuality, both which are in operation here, gesture towards the naturalization of a specific way of seeing: the landscape view that composes and structures the world "so that it may be appropriated by a detached, individual spectator to whom an illusion of order and control is offered through the composition of space according to the certainties of geometry."[28] The invocation of "Heaven" and "guardian angels" is complemented with an attempt to classicize this invented tradition by presenting Britain in its anthropomorphic form, the Athena-like figure of Britannia. By presenting Britain as Britannia, the lyric activates the allegorical where the abstract space of the metropolitan nation is signified as, and through, a concrete figure. Though the figure of Britannia is much older than Thompson's lyric, it nevertheless heralds a key way of representing Britain, especially visually, to its colonies in the succeeding years. Let us take a brief excursus to examine some of these visual depictions.

In 1777, the East India Company commissioned the Italian painter Spiridione Roma to paint an allegorical ceiling piece for the Revenue Committee room in the East India House at London. Completed the next year, the painting titled *The East Offering its Riches to Britannia* (Figure 3.1) depicts a dark-skinned woman representing "India," offering pearls to Britannia, the personification of Britain. The painting also includes in its frame "China" (kneeling behind "India"), awaiting his turn to gift a Ming vase; a personification of the Thames, signifying the power of London; and the lion, symbolizing the British State. In the background is an East India Company ship with a flag bearing stripes, and the St. George's cross. While Roma's painting is by no means the first depiction of Britannia, it nonetheless provides an important instance of the anthropomorphic depiction of Britain in the age of empire. In the wake of the 1857 Uprising in British India, the anthropomorphic image of Britain as well as other kinds of personifications were deployed by British periodicals such as *Punch* to convey notions of British frailty and Indian brutality. These figures were also used to speak of the retribution and justice delivered upon the revolting mutineers by their colonial masters. For instance, the *Punch* cartoon from August 1857 called *The British Lion's Vengeance on the Bengal Tiger* (Figure 3.2) allegorizes the Indian Uprising as a fight between the English lion and the Indian tiger. The lion, which symbolizes the colonial British state is shown attacking the Bengal tiger, representing the Indian mutinous rabble, for attempting to maul an Englishwoman and her child. The woman under the tiger's paw signals both an individual Englishwoman but also evokes the image of a helpless Britannia under attack. And understandably, the English lion must protect her honour.

Figure 3.1 The East Offering its Riches to Britannia, 1778, Spiridione Roma

Source: © The British Library Board (Foster 245).

Figure 3.2 The British Lion's Vengeance on the Bengal Tiger, 1857
Source: John Tenniel. *Punch*.

Another *Punch* cartoon from the following month of 1857 shows a personified Justice in the thick of battle, and delivering justice to the natives. She is shown not only leading British troops against the mutineers but also protecting Indian women from them. The cartoon is implicated in the trope of the colonizer (Justice, in the figure of an Englishwoman) saving native women from native men. And the martial nature of the figure of Justice, though without classical regalia here, recalls Britannia for its readers. There are also instances of the more direct use of the Britannia image. For instance, in 1885, at the height of the Great Game when Britain feared a Russian invasion of their Indian colony, *Punch* published a cartoon titled *Ready!* that depicted Britannia with her "pets," the Lion and the Tiger (signifying India but now clearly "tamed"), waiting to confront the Russian advance into Afghanistan and beyond.

To now return to the discussion of Thomson's lyric: we should note that the trope of personification, though important for the lyric, is not the only way it engages the allegorical. Constructing a distinction between Britain and other spaces, Thomson tells us:

> The nations, not so blest as thee,
> Must, in their turns, to tyrants fall;
> While thou shalt flourish great and free,
> The dread and envy of them all.[29]

The lines construct a distinction between a "great and free" Britain, the "dread and envy of all," and other nations "not so blest" that must in time "to tyrants fall." They construct an allegorical relationship between Britain and freedom, one that also gestures towards the equivalence between other places and unfreedom. The lyric, then, invokes the allegorical in its spatial imagination in two distinct ways: by *idolizing* Britain (as Britannia) and through *idealizing* it (as "great and free" and so on). Its delineation of the qualities of British being are paralleled by a disavowal of historicity. Note how in presenting Britain as "great" and "free," the lines do not acknowledge Britain itself as one of the "tyrants" though the British imperial project in North America, and the attendant expropriation of Native Americans, was well underway during this time. The lyric's disavowal of historicity – which we could think of, following Theodore Adorno, as its *gesture of refusal* – is of critical importance.[30] It is what gives the lyric the ideological charge to glorify the land and its people. It, furthermore, collapses the distinction between British imperialism and nationalism, charting out a historical trajectory for Britain and Britons, differentiating them as singular from other spaces and peoples. This sense of singularity, notably, is structured through the claim to *essential* superiority that can only be made through a refusal of history.

This claim to, and articulation of, an essential difference between metropolitan and peripheral spaces would continue to structure metropolitan patriotic lyrics throughout the imperial era. As British imperialism put down roots and consolidated in South Asia after the loss of Britain's American empire, lyrics written about its Indian empire typically exuded claims to British superiority. This can be seen, for instance, in a hymn from 1819 composed by Reginald Heber, the Anglican Archbishop of Calcutta from 1823 until his death in 1826. Heber's hymn opens with:

> From Greenland's icy mountains, from India's coral strand;
> Where Afric's sunny fountains roll down their golden sand:
> From many an ancient river, from many a palmy plain,
> They call us to deliver their land from error's chain.
> What though the spicy breezes blow soft o'er Ceylon's isle;
> Though every prospect pleases, and only man is vile?
> In vain with lavish kindness the gifts of God are strown;
> The heathen in his blindness bows down to wood and stone.
> Shall *we*, whose souls are lighted with wisdom from on high,
> Shall *we* to those benighted the lamp of life deny?[31]

Like Thomson's lyric, this hymn also brings together the themes of nationalism and imperialism. These are joined with a Christian mission and a civilizing impulse that anchor the claims to British superiority. This is understandable, especially since the hymn's focus is not Britain but the "heathen" lands – Greenland, India, Africa, Ceylon – that are in need of deliverance from "error's chain." This salvific agenda runs through the hymn, which anticipates the civilizing mission's poetical apogee, Rudyard Kipling's "The White Man's Burden." Notice how the hymn deploys "us" and "we" in the lines above to separate out the "heathen in his blindness" from the British Anglicans. The latter are not only elevated owing to their souls being "lighted with wisdom from on high" but also tasked with providing the heathen with the "lamp of life."

While the hymn points out that the "heathen in his blindness bows down to wood and stone," it nevertheless privileges the heathens' lands. Every "prospect" of these lands is pleasant while the inhabitants ("man") are vile. And particularly illuminating is the hymn's spatial imagination of the heathen lands. It invokes the "prospect" – the landscape view – to depict these spaces and follows this up by noting that "the gifts of God" are strewn about the lands with "lavish kindness." This way of looking, as has been noted by several scholars, is quite typical of a colonial perspective on colonized (or to be colonized) lands, which sees the landscape as one of plenitude that has to be taken and used lest it be wasted.[32] The most powerful instance of this vision is, arguably, Thomas Carlyle's "Occasional Discourse on the Negro Question," where he appeals for a return to slavery in the British Empire. Carlyle bases his argument on the Caribbean, suggesting that it was "European heroism" that made the "putrescences and waste savageries arable" allowing the West Indies to "grow a pumpkin for any negro."[33] The genius of Europe, he writes, can be seen in the productivity of the Caribbean lands:

> The West Indies [now] grow pineapples, and sweet fruits, and spices; we hope they will, one day, grow beautiful, heroic human lives too, which is surely the ultimate object they were made for; beautiful souls and brave.[34]

Notice how the colonial improvement of the land, signalled through its produce of pineapples, sweet fruits, and spices in addition to the "pumpkins," blends seamlessly into a promise of the Caribbean producing, in time, "beautiful" and "heroic" humans and souls. As such, Carlyle shows a perfect alignment of human productiveness with morality.

Heber's, though, is a more poetic vision of space, constructing equivalences between specific heathen locales and commodities of imperial trade and plunder. It signifies the heathen through geographical units such as Greenland (presumably a reference to the Greenlandic Inuit) as well as the India, Ceylon (Sri Lanka), and Africa. And it describes them by using the adjectival form of the commodities. We are told of India's "*coral* strands," Africa's "*golden* sand," and Sri Lanka's "*spicy* breeze," where the commodities in question – rubies, gold, and spices – become anchors for the identities of the locales. Unlike Carlyle, however, the hymn does not stress the role of human labour in producing the commodities in question but presents them, instead, as the "gifts of God." In effect, it *pastoralizes* these spaces, and in so doing, partakes in a gesture of refusal that renders invisible both human labour as well as the history of colonial loot and plunder around these commodities. Half a century after Heber's poem, the colonial condition would provoke a heathen from Bengal to indeed bow down to wood and stone, and fashion one of the most durable anti-colonial icons of South Asia: the image of the space of Bengal as a mother and mother goddess.

Vandemataram: structures of filiation

Vandemataram is a hymn to the ethno-linguistic homeland of Bengal imagined as a landscape that is personified as a mother, and a mother goddess.[35] The formal structure of the song is unique in that it uses both Sanskrit and Bengali in its composition.[36] The song opens with the incantatory Sanskrit phrase "vandemataram" that is repeated throughout the text:

1. I revere the Mother! The Mother
   Rich in waters, rich in fruit,
   Cooled by the southern airs,
   Verdant with the harvest fair.[37]

The hymn invokes the Bengal landscape, positing it as an image of plenitude, which is not just "rich in waters, rich in fruit," but also green from the standing crops, and "cooled by the southern airs." It invokes the traditional – and transcultural – literary trope of equating womanhood, nature, and fecundity from the outset, and develops it over the course of the poem. It continues this interplay between landscape, homeland, mother, and mother goddess in the subsequent stanza:

2. The Mother – with nights that thrill
   in the light of the moon,
   Radiant with foliage and flowers in bloom,
   Smiling sweetly, speaking gently,
   Giving joy and gifts in plenty.

The song now presents the homeland-as-landscape as the "Mother" signalling towards the ideas of both the maternal and the divine. She is depicted as adorned with flowers and foliage who is also "Smiling sweetly, speaking gently,/Giving joy and gifts in plenty." In other words, it presents the Bengal homeland-as-landscape as a nurturing space as it sets out an affective register for articulating it. The tenor of the song undergoes a shift from the third stanza. The image of the nurturing mother(land) is transformed into an image of power – indeed, retributive power:

3. Powerless? How so, Mother,
   With the strength of voices fell,
   Seventy millions in their swell!
   And with sharpened swords
   By twice as many hands upheld!

4. To the Mother I bow low,
   To her who wields so great a force,
   To her who saves,
   And drives away the hostile hordes!

The song critiques the notion of the mother(land) as "powerless." Instead, it argues that she gains her strength from her 70 million children who are capable of wielding "Twice as many" swords against any opponent. The deified landscape is presented in the fourth stanza as a source of power: the mother is the saviour who "drives away the hostile hordes." In the stanzas that follow, the image of the personified Bengal homeland-as-landscape as an embodiment of power is also, crucially, matched with its conception as an object of devotion:

5. You our wisdom, you our law,
   You our heart, you our core,
   In our bodies the living force is thine!
6. Mother, you're our strength of arm,
   And in our hearts the loving balm,
   Yours the form we shape in every shrine!

The mother is presented as the source of wisdom and law; she is also projected as the vital essence that animates her children. In a display of extreme affect, the poem claims that the mother is "our heart . . . our core"; she is also imagined as the living force that enlivens the bodies of her children. This affective turn continues in the next stanza where the mother(land) is conceived as the "strength of arm" and "loving balm" of Bengali hearts. While the maternal personification of the homeland-as-landscape implicitly gestures towards the Hindu notion of the mother goddess, the last line of the sixth stanza makes this explicit. The line moves, quite literally, from idealization to idolization by stating that "Yours the form we shape in every shrine."

The explicit identification of the homeland-as-landscape with the mother goddess continues, and is strengthened, in the seventh stanza where the space is equated with certain female members of the Hindu pantheon:

7. For you are Durga, bearer of the tenfold power,
   And wealth's Goddess, dallying on the lotus flower,
   You are Speech, to you I bow,
   To us wisdom you endow.

The Bengal homeland-as-landscape now appears as Durga, an identification that would undoubtedly resonate with Chatterji's Bengali readers as the festival of Durga was – and continues to be – the most important annual festival of Bengal among Bengali Hindus. The song further continues with its apotheosizing impulse. Not only is the homeland-as-landscape an embodiment of "the tenfold power," it also becomes an incarnation of Lakshmi ("wealth's goddess") and Saraswati (the Hindu goddess of "Speech," learning, and "wisdom"). In the concluding stanzas, the hymn further underlines its devotional tenor:

8. I bow to the Goddess Fair,
   Rich in waters, rich in fruit,
   To the Mother,
   Spotless – and beyond compare!

9. I revere the Mother! the Mother
   Darkly green and also true,
   Richly dressed, of joyous face,
   This ever-plenteous land of grace.

The homeland is now idealized as untainted and incomparable; it also repeats the line from the first stanza ("Rich in waters, rich in fruit") to further underscore the idea of the Bengal homeland-as-landscape as a space of plenitude. *Vandemataram* concludes with the image of the mother(land) who is "richly dressed" in verdure; one who is a life-giver, providing for and nourishing her children.

Vandemataram, then, engages in the idealization and idolization of Bengal. Let me first examine how it idealizes space. The Bengal of *Vandemataram* is a land of plenitude that is "Verdant with the harvest fair." Notice how, in idealizing Bengal, the lyric partakes in a logic of allegorizing space that is reminiscent of Thomson's "Rule! Britannia," which spoke of Britain as "great and free." The lyric is presenting Bengal as "great" as well; but instead of positing Bengal's greatness, it does so through a process of negation. The lyric's vision of plenty contrasts starkly with objective reality – indeed, it is a gesture of refusal of the history, and the lived experience, of famines in Bengal through the late-eighteenth and nineteenth to the early-twentieth centuries. The experience of famines was the stark reality for subjects of the British Empire, in South Asia and beyond.[38] The historian B. M. Bhatia notes that under the East India Company's rule between 1765 and 1858, India experienced 12 famines and four severe scarcities; but between 1860 and 1908, famine or scarcities prevailed in some part of the country or the other for 20 of these 49 years of direct British rule.[39] In the decades that followed the first census of British India in 1871, the average life expectancy of Indians never exceeded 25. It was 24.6 in the decade of 1871–1881; rose to 25 for 1881–1891, and dipped as low as 20.1 during 1911–1921.[40] Ira Klein suggests that epidemics of malaria, cholera, and plague through the late-nineteenth century claimed further lives among British India's colonial subjects. The social reality of famine and epidemics framed the lyric's appearance in the literary marketplace in 1875 (in the journal *Bongodorshon*) and 1882, as part of *Anandamath*; and as a stand-alone song in 1896, when it was first performed by Tagore in the Calcutta session of the

Congress. *Vandemataram*'s affective appeal – and utopian charge – would lie in its presentation of the space of Bengal as a divine and nurturing mother, creating a positive dissonance between the lyric's spatial imaginary and the lived reality of that space as one of distress, marked by death and devastation. By idealizing the space of Bengal thus, the lyric partakes in a determinate negation of the colonial present.

But the lyric's idealization of Bengal has its pitfalls. *Vandemataram* signals the space as a riparian land of fecundity and plenitude; as a spatial view or *landscape* that is replete with trees, flowers, fruits, and standing crops. Implicit in such a conception of the land is that its plenitude is "natural" and a voluntary gift of the mother(land) and not a product of human labour. The only humans present in the landscape are the "children" of the mother(land) who appear as warriors and devotees, with no role in producing the plenitude and fecundity of the space of Bengal. A working country, Raymond Williams reminds us "is hardly ever a landscape."[41] In *Vandemataram*'s spatial imaginary, the space of Bengal is *landscaped*, that is, evacuated of farmers and members of the labouring class.[42] In so doing, the lyric engages in a representational maneuver that is analogous to lyrics from the European pastoral tradition. As Williams tells us, the English country house poems draw on the classical trope of *locus amoenus* poems to mystify the social division of labour. For this, it harnesses the Christian myth of the Fall, and magically removes the "the curse of labor" through "a simple extraction of the existence of laborers."[43] *Vandemataram*'s evacuation of all traces of human labour from the space is then an attempt to *pastoralize* Bengal, which was overwhelmingly the land of peasant cultivators.[44] The lyric, then, not only produces the Bengal homeland-as-landscape as a spatial fetish but also attempts to construct the hegemony of the landholding class of Bengalis, that is, the minority landed gentry engendered by the British through the introduction of the "rule of property" in Bengal, and beyond.[45]

The lyric's idealization of the mother(land)'s power is similarly ambivalent. *Vandemataram* claims that the mother gains her strength from her 70 million children who raise a clamour [সপ্তকোটিকণ্ঠকলকলনিনাদকরালে; *shopto-koti-kontho-kolo-kolo-ninad-koraale*]; the children are able to wield twice that number of swords [দ্বিসপ্তকোটিভুজৈধৃতখরকরবালে; *dwishopto-koti-bhujoidhrito-khoro-koro-baale*]. This is a vision of strength where, as Sudipta Kaviraj notes, the "single destinies are erased . . . [and the mother's children] are able to see themselves as a collective maker of a single enunciation."[46] But this "single enunciation" that counters the colonial depredations of the *mother*land is also marked by a parochial politics about which Kaviraj remains peculiarly incurious. To illuminate this, we need to gloss the lyric's use of the fairly exact figure of 70 million – the actual word used is *shopto-koti* [সপ্তকোটি] which translates as "seven crores." Sugata Bose tells us that the "magic number of seven crores" refers to Bengalis but that figure needs further explication.[47] The figure appears to come from the colonial censuses that show the population of the Bengal province ranging from 62.6 million in 1871 to 69.5 million in 1881.[48] The 1871 census also remarks that Hindus were about

64.5 per cent of the population of the provinces of Bengal and Assam. But it goes on to note that:

> of the 20½ millions of Mussulmans in Bengal and Assam (forming the larger moiety of the Mahomedan population of British India), 17½ millions are found in Eastern Bengal and the adjoining Districts of Sylhet and Cachar, where they amount to 49 per cent, of the total population; and in two districts, those of Bogra and Rajshahye [Rajshahi], to about 80 per cent. In that part of the country they comprise the bulk of the cultivating and labouring class.[49]

We must remember here that the British Indian province of Bengal comprised what is today the nation-state of Bangladesh and Indian provinces of West Bengal, Bihar, Odisha, and Jharkhand. The regions of Bihar, Odisha, and Jharkhand, which made up the bulk of the Hindu population of the province, however, were not part of the territorial identity of Bengal. While presenting Hindus as the majority religious community of the Bengal province, the census nevertheless unequivocally acknowledges that Muslims constituted the majority of Bengal proper. The lyric's counter-colonial imagination of collectivity and power that presents the Bengal homeland as a mother goddess, then, produces a Hindu conception of the space of Bengal that makes its Muslim population invisible. The politics of colonial enumeration also treated the lower castes and the tribal people – the Bengali word for them is *adibashi* [আদিবাসী] that translates as "original residents" – as part of the Hindu fold, even though they were not considered equals by upper-caste Hindus.[50] The invocation of the "seven crore" then gestures towards the construction of a hegemonic Bengali identity in the colonial moment, one that was decidedly Hindu of an upper-class and upper-caste provenance. The consolidation of – and resistance against – this minority Bengali identity as hegemonic would inform the contentious politics of colonial Bengal, eventually leading to its partition in 1947.[51]

Besides idealizing it, *Vandemataram* also *idolizes* the space of Bengal. It does so by presenting the space as a manifestation of the mother goddess of the Hindu pantheon drawing, in particular, on the iconographic representation of Durga as well as aspects of Hindu liturgy. In addition, *Vandemataram* is also *hymn* to the idealized and idolized space of Bengal. It gestures towards the emergence of a new literary genre across South Asia that develops out of extant forms of Sanskrit (and Hindu) devotional poetry and hymns. In Bengal, these group of songs came to be called *deshatmobodhak gaan* [দেশাত্মবোধক গান] literally meaning "songs that raises self-awareness about the country" or, to use my descriptor for them, the patriotic lyric.[52] Its use of the figure of Durga, and its reliance on Hindu liturgical practices and devotional forms show the *Vandemataram*'s reliance on the conceptual and institutional frameworks as well as the imaginative resources of Hinduism. In addition, it points to the move across British India in the late-nineteenth century of using a selective array of Hindu discourses and rites to not only define, but also present a "compelling image of a grievously wronged community that demanded retribution for past wrongs and transgressions."[53] Furthermore, these efforts posited Hinduism as the civil religion of India in what has been termed the "secularization of Hinduism."[54]

The lyric's use of the figure of the Durga is particularly significant. For one, it resonates with the increasingly public nature of the annual Durga festival in Bengal, especially Calcutta, from the mid-nineteenth century. The festival celebrates her victory over, and killing of, the male warrior-king Mohishashur who is typically depicted as a demon.[55] In Bengal, she is popularly understood as a married daughter from the region, and the annual festival in her honour is understood as the time she visits her natal home (Bengal) from her husband's (the Himalaya). This is also the dominant feeling evoked in the *agomoni* [আগমনী] and *bijoya* [বিজয়া] songs, the corpus of traditional Bengali devotional sung lyrics to Durga that commemorate, respectively, her annual arrival to, and departure from Bengal. In addition, she is also shown visiting Bengal with her two daughters and two sons and is worshipped *en famille* with her children. The lyric's spatial conception of Bengal as a mother goddess enables it to fashion an image of space that can be, and is, used to counter allegorical representations of the colony by the British (for instance, in the painting by Roma and the *Punch* cartoons). More crucially, it enables the lyric to yoke together the spiritual and the spatial, which it does not just through its use of the image of Hindu divinity, but also by appropriating particular aspects of Hindu liturgy.

The system of beliefs and rituals that comprise Hinduism moves from the aniconic world of the Vedas to the world of the epics and Puranas marked by the profusion of idols. A key aspect of the latter is the service and adoration for the divine image, that is, a physical image that represents the divine. The Sanskrit language has two words for the divine image through two words: *pratima* [प्रतिमा], which refers to "the image as replica, portrait or reflection," and *murti* [मूर्ति], signifying a "corporeal form enclosed or frozen within a set limit, a condensed presence, and arrested movement."[56] Hindu ritual understands the idols not as the divine itself, but a manifestation of the infinite and formless divinity in finite form. As Charles Malamoud tells us, each of these terms indicates "a certain distance between the idol and the god: *pratima* is a copy of an inaccessible original, whereas *murti* signifies a matrix for the concretization of a divine fluidity that cannot be grasped."[57] For the idols to be worshipped as divinity, they must first be ritually transformed, and given life. Liturgically, this translates into specific rites, namely, that of "opening the eyes" [अक्षिउन्मेशन; *askhy-unmeshana*] and "installation of the breaths" [प्राणप्रतिष्ठा; *prana-pratistha*] through which the inanimate idol is animated.[58]

Vandemataram is written in Bengali, a vernacular of eastern India, and Sanskrit, the language of Hindu liturgy that enjoys a status similar to, say, Latin in the Roman Catholic church. The study of Sanskrit and the sacred texts written in that language have traditionally been the exclusive preserve of the sacerdotal caste, the brahmins, considered the highest among the Hindu castes who, as priests, also perform the religious rituals. The apotheosizing impulse of *Vandemataram* that idolizes and deifies the land – the Bengal landscape – into the mother goddess brings the landscape to life, as it were, and in so doing replicates this liturgical trajectory. This is further underlined by the lyric's reliance on Sanskrit, as well as the incantatory repetition of the phrase "vandemataram," which mimics ritualistic utterance. The lyric could potentially be sung by any of the "mother's children," and thus lays claim to a democratization of caste-specific sacerdotal functions. The quasi-egalitarian claim of this move, however, goes hand in hand with the normalization of the sacerdotal functions *per se*, and of the attendant priestly class. The lyric replicates the extant caste hierarchy of Hinduism, if in a more nuanced manner, and attempts to reproduce the hegemony of the brahminical caste.

Chatterji's formal innovation lies in not just fashioning a modern form (the patriotic lyric) but in creatively reworking extant forms of Hindu religious panegyrics, the *stuti* [स्तुति], the *stava* [स्तव], or *stotra* [स्तोत्र] for a distinctly non-religious purpose. This transformation, in Kaviraj's formulation, is twofold: it moves the essentially personal nature of the traditional *stotra* to a collective utterance of a political anthem while also displacing the sacred on to the secular to construct a "secular sacredness."[59] Chatterji's appropriation of the formal structure of the *stotra* in *Vandemataram*, as well as his use of Sanskrit in tandem with Bengali is crucial and appropriate. The Sanskrit language provided "cultural ballast" and "legitimising authority" to the lyric while the use of the Sanskritic form "lent it gravitas."[60] This is of course coupled with Chatterji's use of Bengali, the "sense and sensibilities" of which allowed for it to develop, in Lipner's words, an "emotional grip" on its readers and auditors.[61] In effect, the lyric's use of both Bengali and Sanskrit allowed it to achieve the "best of both worlds – the authority of tradition and the enveloping freshness of current speech."[62]

The single-most important impact of *Vandemataram* is its illumination of a specific spatial desire: the inauguration of a new conception of space where the Bengal landscape is represented as, and through, the mother goddess. This sets up the figure as the mediating category that allows for the construction of an equivalence between a concrete *place* (the local landscape) and the abstract *space* (Bengal, and subsequently, India). The figure of the mother goddess provides an experiential register to abstract space, and thus aids in its transformation, and translation, into concrete place. But even as the lyric recodes the abstract as the concrete, it infuses the mundane – the landscape – with the spiritual, in effect, *re-enchanting* the disenchanted space of colonial capitalist modernity. The spatial imagination of *Vandemataram*, moreover, relies exclusively on the Hindu conceptual universe, which in turn guarantees that imagination. The lyric's idolization of the Bengal landscape *spatializes* Hinduism. This conceptual maneuver simultaneously renders the space into a Hindu entity: it *Hinduizes* the space of Bengal. This process is structured by the lyric's generation of a productive misrecognition where notions of devotion and piety associated with ordinary lived Hinduism are displaced, in a catachrestic manner, on to the space of Bengal (and by extension, the space under the sign of "India"). In such a scheme, the inhabitant of the space (of Bengal or India) is interpellated as a devotee, whose devotion for – or ways of perceiving, conceiving, and being in – that *space* is sanctioned and validated by the *Hindu* frameworks of faith and piety. In so doing, the lyric sets up an exclusive notion of space that can only be accessed from within the Hindu fold. To put this somewhat differently, *Vandemataram* sets up the categories of inclusion and exclusion on the questions of belonging: who belongs to the space? And to whom does the space belong? If *Anandamath* excluded Muslims explicitly from the space of Bengal, the song's ideological labour is more implicit. By excluding all non-Hindus ways of partaking in the imagination of the space of Bengal and, by extension, India, it excludes non-Hindus from those spaces.

The representational protocols that *Vandemataram* introduced would be normalized and become reified into tropes of spatial representation in due course. These tropes were used to great effect by Tagore in his poem "Bongomata" [বঙ্গমাতা; "Mother Bengal"] that stands in a parodic relation to *Vandemataram*. Published in 1896, the same year he set Chatterji's lyric to music, the poem reads as both a critique of, and a plea to, the Bengal mother(land):

> Through ups and downs, in virtue and in sin
> Let them be humans, all your offspring.
> O lovestruck Bengal, from your lap let go
> These eternal babes, so they may grow . . .
> O awestruck mother, your seven crore children
> You've kept them Bengali, not made them human.[63]

Tagore's poem uses several recognizable protocols of *Vandemataram*: the anthropomorphic (though not deified) image of space as a mother, the inhabitants of the Bengal homeland as her "children," and the specific number of seven crores. But Tagore's use of these tropes is accompanied by a powerful reversal of their connotations. If Chatterji's lyric posited a mother made strong by her seven crore children, Tagore offers a rebuttal to that hyperbolic vision. He deflates the heroic tenor of *Vandemataram* by proffering a literal rendition of the idea of the mother's children. They are, he suggests, precisely what Chatterji says they are – eternal babes [চিরশিশু; *chiroshishu*] who are yet to venture out of their mother's lap, and home. This is coupled with a stinging rebuke: Mother Bengal's children have been made to remain Bengali and not been allowed to grow into human beings.

The popularity of *Vandemataram* would lead to the wide dissemination of the genre across British India. Over time, the spatial references in *Vandemataram* came to signify the British Indian colony instead of Bengal, and other patriotic lyrics would draw on it to write about British Indian space in an anthropomorphic and apotheosizing vein. A key example is provided by Dwijendralal Ray, a junior contemporary of Chatterji whose reputation in the Bengali literary universe rests on his corpus of patriotic lyrics besides his historical plays. Note, for instance, the opening lines from his patriotic lyric titled "Bharotborsho" [ভারতবর্ষ; "India"]:

> The day you rose, Mother India, from the blue ocean,
> The world, mother, cried out in joy, in sheer devotion![64]

The song opens with the image of clamour, joy, and devotion at the sight of Mother India emerging from the blue ocean. The lyric's chorus line goes on to posit India as the "enchantress of the world" [জগন্মোহিনী; *jagonmohini*], and the "mother of the universe" [জগজ্জনী; *jagojjonini*], terms typically used to describe Hindu mother goddesses. The opening, moreover, offers a striking parallel to that of Thomson's "Rule! Britannia." The anthropomorphism of space, the transcendental perspective on it, as well its supposedly divine origins all have a place in Ray's lyric with the figure of Mother India replacing Britannia. The peripheral lyric in Ray's hands departs from its metropolitan counterpart in another decisive way. The latter's celebration of imperial power in the chorus – the exhortation to Britannia to rule the waves – in the peripheral context becomes recoded into the devotional vein and is invested with a spiritual charge through the catachrestic use of terms such as *jagojjonini*. Note the subtle difference between the ways in which Ray and Chatterji deify the spaces in their lyrics. Chatterji, as we have seen, presents Bengal as Durga, Lakshmi, or Saraswati. This is done through Chatterji's use of the Sanskritic phrase "you are" [ত্বম হি; "*twam hi*"] that effects this equivalence [ত্বম হি দুর্গাদশপ্রহরণধারিনী/ কমলাকমলদলবিহারিনী/ বাণী বিদ্যাদায়িনী; *twam hi Durga dashapraharanadharini/ kamala kamaladala biharini/ vani vidyadayini*]. Ray is not suggesting an explicit equivalence but presenting the India *as* a distinct goddess and a mother by referring to India [ভারতবর্ষ; *Bharotborsho*] in both registers. India is both mother [জননী; *jononi*] and mother of the universe [জগজ্জনী; *jagojjonini*]. The lyric's opening signals the sedimentation of the image of deified space – Mother India – as a representational trope. The patriotic lyric has well and truly emerged in its Indian variant. But this would not be the only, or only kind of, innovation with the form of the peripheral patriotic lyric. Tagore's *Jana Gana Mana* would offer another instance of innovation with this literary form.

Jana Gana Mana: structures of affiliation

Tagore's *Jana Gana Mana* is a patriotic lyric about the space of India. And like other members of the patriotic lyric genre, it is also fashioned as a *hymn* that is meant to be sung. This finds expression in Tagore's title for the lyric's English translation, which he calls the "Morning Song of India."[65] The lyric is marked with a spatial desire that seeks to valorize the colonial space of India. But the conceptions of space and spatial belonging that it presents are not just different but fundamentally antithetical to the spatial imaginary of Chatterji's *Vandemataram*. Tagore's lyric opens with:

> Guide of people's thoughts, India's destiny you decree.
> Glory be to thee!
> The Punjab, Sindh, Gujarat, Maratha
> The Dravid peninsula, Odisha and Bangla;
> The Himalayan peaks, and the Vindhya,
> The surging waves of the Yamuna and Ganga –
> They rise in your name, seek to be blessed,
> Singing you hymns of praise.
> For you the people's well-being – glory to thee!

India's destiny you decree,
Glory, glory, glory to thee![66]

Tagore's use of the compound form of *bidhata* [বিধাতা] signals his invocation of the divine as the word, literally meaning someone who decrees, is typically used to refer to god. The hymn is addressed to *bharotobhaggyobidhata* [ভারতভাগ্যবিধাতা] or one who decrees India's destiny. The devotional tenor of the song is further underlined through the refrain – "glory to thee!/India's destiny you decree,/Glory, glory, glory to thee!" [জয় হে জয় হে জয় হে, জয় জয় জয় হে] – that is used to conclude each stanza of the lyric. And significantly, the conception of the divine here is not a specific one moored to this or that religious tradition but of a more general conception. As will become evident from my discussion of it, such a conception of the divine will inform the lyric's distinctive spatial imaginary.

As the lyric delineates India, it employs a certain restlessness of vision and an insistent enumerative protocol. The restlessness I speak of can be seen in the way India is signified throughout the lyric. It is signified through markers of its political and physical geographies in the first stanzas; through religious communities in the second, before gesturing to India through historical time. The lyric opens by presenting India in an enumerative vein; as a concatenation of the various ethnolinguistic homelands of South Asia, namely, Punjab, Sindh, Gujarat, the Maratha lands, the Dravidian peninsula, Odisha, and Bengal. India, significantly enough, is not signified as a space of uniformity but as an entity that is at once composite and variegated: as a civilizational space that anchors these distinct *patrias*, to use Chris Bayly's term for the homelands.[67] This is paralleled through the lyric's signalling of India's physical geography, evoked through the rivers Ganga and Yamuna in addition to the mountain ranges Himalaya and Vindhya. A cartographic view of South Asian space is evident in both these articulations since the political and physical geographies of South Asia are presented as if laid out as spatial views, that is, as landscapes. These physical and political landscapes are depicted as chanting the name of the divine who rules Indian minds, and decrees India's destiny. Tagore's deployment of the spiritual is, crucially, different from Chatterji or even Ray. India is not equated with gods, and neither is the space deified; instead, Tagore makes, and maintains, a distinction between divinity and India throughout the lyric.

The poem's next stanza continues the strategy of enumeration but now focuses on the various communities that inhabit India. Signalling the space of India as a fundamentally open one, the lyric notes:

Hearing your welcome, heeding your call,
Hindu, Sikh, Jain and Buddhist, they all
With Muslim, Christian, Parsee meet,
To weave love's garland by your seat.
For you the people's unity – glory to thee!
India's destiny you decree,
Glory, glory, glory to thee!

The divinity who controls the destiny of the land and its people is now shown as welcoming diverse communities into the space signified as India. The call of this voice brings together, in the lyric's imagination, Hindus, Muslims, Sikhs, Buddhists, Jains, Parsees, and Christians to the deity's shrine where they weave "a garland of love." This stanza seeks to construct, most directly, an idea of an Indian community that is marked by a sense of affiliation, and not filiation. In other words, the community is constituted by the forces of history and not structured by the relations of blood or soil. In so doing, it makes visible the multiple religious communities that inhabit South Asia, situating them as groups who belong, and are equal claimants, to Indian space. This stanza of *Jana Gana Mana* exudes a spatial desire of a very specific kind, one that presents the space of India as fundamentally open and welcoming. This would be a critical aspect of Tagore's idea of what India has been, is, and should be – in other words, his idea of India in the pre-colonial and colonial eras, and as a postcolony. This is an idea he would revisit throughout his life and which would find expression in his poems, lyrics, plays, and novels such as *Gora* (that I discuss in the next chapter) but also in his many lectures and essays.

The idea of the open and welcoming India returns forcefully in another of Tagore's patriotic lyrics titled "bharot-tirtho" [ভারততীর্থ] that is typically translated as "Pilgrimage to India."[68] There he writes:

> No one knows at whose great call
> Streams of humanity
> In a mighty tide flowed who knows whence
> To mingle in that sea.
> Aryan and non-Aryan came,
> Chinese, Dravidian,
> Scythian, Hun, Mughal, Pathan
> In body blent as one:
> And now the West unfolds its doors,
> The world bears bounty from its store –
> Give and receive, merge and be merged:
> None will excluded be
> From India's ocean-shore of great humanity.[69]

Tagore speaks of India as an ocean of humanity that is composed of its various communities. Jahan Ramazani has noted that anti-colonial poets often signify the "sublimity of decolonization" by gesturing towards "cultural and national multiplicity one squelched by the homogenizing force of empire."[70] While both *Jana Gana Mana* and "bharot-tirtho" underline the cultural and national multiplicity of Indian space, they nevertheless also signal a concomitant unity, one that is distinct from the homogenizing squelch produced by colonialism. Note how "bharot-tirtho" speaks of the community of Indians as people who come from beyond what is typically conceived of as India. The enumerative modality is deployed, as in *Jana Gana Mana*, to make visible the various communities – the "Chinese,"

"Scythian," "Hun," "Mughal," "Pathan" – who are nevertheless "in body blent as one" [এক দেহে হল লীন]. Indian space is signalled as the anchor of these specificities that also, simultaneously, produces a coherence. This coherence is not simply an idea of "unity in diversity" but one where the whole is greater than the sum of its parts, thus suggesting a kind of dialectical unity of the parts and the whole. It is also tempting to read this articulation of Indian space as a radical reworking of the idea of colonial progressive history that the British introduced in India. Instead of reproducing the colonial teleology of rule that presented India as a space ruled successively by Hindus, Muslims, and eventually the British, the lyric re-interprets that logic to suggest India as a singular site of global processes of interaction between peoples, faiths, and ideas.

The people who come into the space of India, it suggests, become a community of Indians. This idea is further stressed through the lines, "Give and receive, merge and be merged:/None will excluded be/From India's ocean-shore of great humanity" [দিবে আর নিবে, মিলাবে মিলিবে, যাবে না ফিরে – /এই ভারতের মহামানবের সাগরতীরে]. The idea of India as an ocean evokes the image of the particular and the collective simultaneously, gesturing towards the individual ethnic and religious communities coming together to form an expansive geo-body. Belonging to this space here is, once again, not structured by the fact of filiation but through an act of affiliation. The spatial desire of this lyric is one that seeks to transcend the bounds of the region and colony quite decisively, gesturing towards what can be only described as a truly radical internationalism. It seeks to make the world its home, and not allow the parochialism of petty identity to imprison one's conception of the world. To say this in another way, the spatial desire of this lyric indicates a community where the foreign is made not just *familiar* but also – and just as crucially – *familial*. In so doing, it also conversely *defamiliarizes* the familial, expanding familial relations to the scale of the global even as it brings the global into the realm of the local. This move – and Tagore will narrativize this in *Gora* – radically negates the logic of filiation to make it utterly superfluous. And this positing of India as an open and welcoming space stretches the idea of India to the point where it becomes a metonym for the world, illuminating a spatial desire that is marked by a utopian politics. And these articulations of spatial utopianism proffer an expansive, and inclusive, conception of the postcolony, locating these in opposition to the parochial and limiting spatial imaginary of Chatterji's *Vandemataram*.

The lyric's vision of affiliation is not only restricted to a horizontal comradeship that seeks to transcend communal divisions. It also takes up, and critiques, extant hierarchies of caste thereby inscribing his spatial imagination with another utopian horizon. Tagore writes:

> Aryan and non-Aryan come,
> Hindu and Musulman:
> Come, O Christian; and today
> Come, all you Englishman.
> Come, Brahman, with a heart made pure
> Hold hands with one and all:

Come, you outcaste: let your load
Of insult from you fall.
To the mother's coronation haste:
The sacred pitcher yet awaits
It's holy water, by the touch
Of all lent sanctity
On India's ocean-shore of great humanity.[71]

The first two lines of this stanza take up the major *horizontal* divisions within Indian society, that of the Aryan and non-Aryan, that is, the ethnic distinction between North and South Indians. Notice also how the lyric includes the English in its imaginative geography, thus seeking to transcend the colonizer–colonized divide. But most crucially, the lyric takes up caste divisions within Hindu society that are typically ignored by upper-class, upper-caste writers from colonial India. The lyric calls on the upper castes, signified through the figure of the brahmin [ব্রাহ্মণ; "Brahman"] to join in "with the heart made pure" [শুচি করি মন; *shuchi kori mon*]. Let them be very clear lest there be any misunderstanding on this point: the lyric is not suggesting that the brahmin is of pure heart; instead, it points to the need for the brahmin to make his heart pure. By pointing to the need for the supposedly upper castes to purify their hearts, the lyric offers a trenchant critique of the ideology of caste. It also speaks to the outcaste using the word "potit" [পতিত] which translates to "fallen" or "low," which is similar to the word "dalit" that is used by the so-called lower castes in the contemporary moment. While the lyric imagines an equitable society with the dalit or outcaste without their "load/of insult" [হোক অপনীত সব অপমানভার], it also exhorts the brahmin to "Hold hands with one and all" [ধরো হাত সবাকার]. This indicates a rejection, and an implicit criticism, of the practice of untouchability that was, and continues to be, practised by upper-caste Hindus.

This concluding stanza from "bharot-tirtho" is also similar to *Jana Gana Mana* on another account. The poem concludes with the call for all communities to sanctify the pitcher for the mother's coronation through their touch.[72] This invocation of the tactile register reiterates the critique of untouchability besides valorizing the contact between communities, ethnicities, and castes. The tactile, moreover, allegorizes labour, inscribing it as a key ingredient in the socio-spatial life of India. Recall the lines from *Jana Gana Mana* where the different communities are shown as weaving "love's garland" by the seat of the divine who decrees India's destiny. These poetic invocations of labour at the heart of the lyrics set up India as a common and evolving project that is the product of a collective, and directed, use of labour. They show the relationship between space, communities, and belonging, as active processes that are structured by, and through, social agents – human beings – whose agency is not mystified. In foregrounding the social relations that make up India as a lived, and living, space, the lyrics move away from mystifying tendencies of the pastoral mode that we have seen in *Vandemataram*. In other words, these instances suggest Tagore's patriotic lyrics seek to present Indian space in a *counter-pastoral* mode.

The privileging of the counter-pastoral in the lyric necessarily engenders a defetishized rendition of space. In other words, the lyric gestures towards space in the process of becoming, that is, as produced and reproduced through historical processes. History is directly adduced in the remaining stanzas of *Jana Gana Mana* through the use of the trope of Time [যুগ; "joog"] and the eternal rider of the chariot:

Through fall and rise, on Time's rugged way
Your chariot-wheels ring, night and day
In that mighty clamor, O eternal rider,
The sound of your conch-shell we hear.
From pain and distress, our savior!
You direct the people's paths – glory to thee!
India's destiny you decree,
Glory, glory, glory to thee!

India's destiny, the lyric notes, is connected to the passage of time. The divine is presented in an anthropomorphic vein, as one who rides Time's chariot [চিরসারথি; *chirosharothi* or the "eternal charioteer" – "eternal rider," in my translation]. This rider is presumably male, and his conch-shell is heard amidst the "mighty clamor" of history. The next stanza continues the depiction of India in a historicizing vein:

Through the nights, dark and still
When this swooning land was ill
We had your lap, through visions and our terror
You watched and blessed us, our loving mother!
You quell the people's sorrow – glory to thee!
India's destiny you decree,
Glory, glory, glory to thee!

The lyric elaborates on the reference to time in the previous stanza through an invocation of the "past." The past is signified through the metaphors of night and the illness. India, it notes, passed through the "dark and still" nights and through a period of illness through the grace of the divine. The divine is now spoken of in another (gendered) anthropomorphic form, as a "loving mother." It is crucial to note, however, that this lyric portrays the divine as being *like a mother* and not as a mother goddess.

The final stanza moves from the idea of an afflicted past to present an image of a hopeful future. The ideas of hope and futurity are presented through images of daybreak, singing birds, and the blissful air. Though it does not speak of decolonization directly, it is quite plausible to think of the dawn as a utopian horizon of counter-colonial possibility. The final stanza of the lyric reads:

Now night gives way to the break of day,
The sun rays fall on the eastern hill

The singing birds, the blissful air,
Life with your joy and mercy, they fill
India awakes, bowing to your feet,
Glory to thee, O King of kings
India's destiny you decree,
Glory, glory, glory to thee!

Tagore's use of the figure of the sun here is not coincidental. His name *rabindra* [রবীন্দ্র] literally means "the Lord of the sun." Its diminutive form *robi* [রবি] is a Bengali word for the sun. By incorporating the sun in the final stanza, Tagore is inscribing himself allusively on to his lyric. This, however, is not an idiosyncratic act of egotism. Through this act of inscription, he is in fact partaking in an important literary device that can be found in poetic traditions across South Asian languages. This device, called *bhonita* [ভনিতা] in Bengali, allows poets to insert themselves at the end of their poems and function as literary signatures. These can often take the form of direct references to the poet's name at the end of a composition.[73] In other instances, the name of the poet is alluded to in an indirect manner and in a more poetic register.[74] The use of the image of the rising sun in this stanza, then, allows Tagore to achieve two ends: it allows him to "sign off" while also using the image symbolically, making it stand in as the marker for a utopian temporal horizon.

Note how the lyric speaks of the divine in another anthropomorphic register from the two previous ones. The divine, which was called the "eternal rider" [চিরসারথি] and "mother" [মাতা] earlier, is now signalled through the phrase "King of kings" [রাজেশ্বর]. Tagore's reference to the divine as the "King of kings" resulted in a controversy during the visit of British king George V to India in 1911. At the INC session held in Calcutta to commemorate the royal visit, *Jana Gana Mana* was sung at the start of the second day's proceedings after which a resolution was passed welcoming the royal couple and expressing loyalty to them. The English newspapers *The Statesman* and the *Englishman* reported that Tagore's *Jana Gana Mana* was a panegyric composed in honour of the British (then also Indian) king.[75] Tagore of course refuted this charge forcefully, writing in a letter:

> A friend, influential in government circles, had importuned me to compose a song in praise of the King. His request had amazed me, and the amazement was mingled with anger. It was under the stress of this violent reaction that I proclaimed . . . the victory of the Dispenser of India's destiny . . . who dwells within the heart of man and leads the multitudes. That the Great Charioteer of Man's destiny in age after age could not by any means be George the Fifth or George the Sixth or any other George, even my loyal friend realized.[76]

In another letter written two years later on the same issue, Tagore would once again repudiate this slur, writing:

> I should only insult myself if I cared to answer those who consider me capable of such unbounded stupidity as to sing in praise of George the Fourth or

> George the Fifth as the Eternal Charioteer leading the pilgrims on their journey to countless ages of the timeless history of mankind.[77]

While the poet's rebuttals undoubtedly settle the issue, the confusion at the heart of the controversy itself points to a key aspect of the lyric: a fundamental ambivalence regarding the depiction of the divine and the space of India.

Throughout the lyric, Tagore signals the divine through a range of representational strategies. We have already seen his reference to the divine as the one who decrees India's destiny [ভারতভাগ্যবিধাতা; *bharotobhaggyobidhata*] and, in a more anthropomorphic vein, as the "eternal rider" [চিরসারথি; *chirosharothi*], "mother" [মাতা; *mata*], and the "King of kings" [রাজেশ্বর; *rajeshwor*]. In addition, the lyric also refers to the divine through its attributes, calling the one who ordains the "people's well-being" [জনগণমঙ্গলদায়ক; *jonogono-mongolodayok*] and "people's unity" [জনগণঐক্যবিধায়ক; *jonogono-oikkyobidhayok*]; one who directs the "people's paths" [জনগণপথপরিচায়ক; *jonogono-pothoporichayok*] and quells the "people's sorrow" [জনগণদুঃখত্রায়ক; *jonogono-dookkhotrayok*]. The lyric employs both religious as well as non-religious designations for the divine. But more significantly, the religious designations are not reducible to any one specific religious tradition. For instance, speaking of the divine as the "mother" and the "King of kings" resonate with both Hindu and Christian conceptions of the spiritual. Similarly, the reference to the divine through its attributes gestures towards the way Islam articulates the divine. By positing several competing, and at times, incompatible, depictions of the divine, the lyric eschews any stable depiction of the category. This also brings together the iconic and aniconic registers of religious representations, thus combining aspects of cataphatic and apophatic theologies in a single text.[78] This is not surprising as Tagore belonged to the Brahmo Samaj, the reform movement and sect within Hinduism that, like Islam, refer to the divine in an aniconic manner and solely through its attributes. The eschewing of a stable or singular notion of the divine then allows for a plural conception of divinity. Notice also that the plural conception of divinity is anchored in the lyric's unitary poetic address. By making a *singular* voice anchor the *diverse* ways of addressing the divine, the lyric activates a heteroglossic mode of address that is crucial – indeed essential – for registering the diverse modes of conceiving the divine in the multi-ethnic, multi-religious, multi-lingual space that is India.[79]

It is crucial to note that the lyric's strategy for describing the divine and the space of India are quite analogous. The restlessness of vision and an enumerative protocol that I noted earlier in the lyric's description of Indian space are also applicable to the way it represents the divine. I have also suggested that *Jana Gana Mana* pluralizes the category of the spiritual through its protocols of representations thereby allowing for it to be imagined in more capacious ways. This is also true of the lyric's spatial imaginary. By signifying Indian space, variously, through its political geography, physical landscape, religious communities, and as a space entwined in the larger historical processes that make it go "Through fall and rise, on Time's rugged way" [পতন অভ্যুদয় বন্ধুর পন্থা]. Thus, the lyric eschews a stable and unitary conception of Indian space. And as in the case of conceiving the divine, also deploys a heteroglossic mode of address to signify Indian space that is a contradictory unity of the one and many, produced through dynamic interactions and interpenetrations of the whole and its parts. Significantly, then, the lyric's heteroglossic mode of address illuminates the difficulty – impossibility even – of speaking about the imaginative geography that is India with any degree of singularity. *Jana Gana Mana*, we might then say, dramatizes the utopian search for a common idiom to represent Indian space. Yet, this impossibility of depicting Indian space with any degree of singularity is precisely what renders Indian space singular. It makes it a metonym of the world as such, and of a utopian imaginative geography where the world is simultaneously one and many.

Scholars of Indian anti-colonial nationalism, most notably Partha Chatterjee, have spoken of the importance of the spiritual in any conception of the Indian nation.[80] But if these studies provide important insights into the ideological mechanisms of anti-colonial nationalism, I fear the conception of the spiritual that they bring into their discussions is quite one-sided, and decidedly impoverished.[81] As my engagement with Chatterji's *Vandemataram* and Tagore's *Jana Gana Mana* show, the category of the "spiritual" ranges from the intensely parochial to the radically universal. In so doing, I have illuminated the category through the register of *difference* inscribing it with content and specificity, rescuing it from ahistorical sameness. Likewise, my considerations on the patriotic lyric as a global genre show that the "spiritual" is also a trans-regional, if trans-regionally varied, category, and not the sole preserve of India or even the non-west.[82] This is crucial to remember, not least to prevent deliberations on the metropole and the colony – and, by extension, the colonizer and the colonized – to descend into a simple, and simplistic, binary of the material "West" and the spiritual "Rest."[83] After all, it would be churlish to refuse William Blake and his Christianity the category of the spiritual in these oft-cited lines:

> And did those feet in ancient time,
> Walk upon England's mountains green:
> And was the holy Lamb of God,
> On Englands pleasant pastures seen! . . .
> I will not cease from Mental Fight,
> Nor shall my Sword sleep in my hand:
> Till we have built Jerusalem,
> In Englands green & pleasant Land.[84]

Notes

1 Rajendra Prasad, "Volume 12. 24th January, 1950," [12.167.2]. *Constituent Assembly of India Debates (Proceedings) – Volume XII*. Accessed November 15, 2018 from http://cadindia.clpr.org.in/constitution_assembly_debates/volume/12/1950-01-24

2 The Republic of India's National Anthem is the first stanza of *Jana Gana Mana*. It recognizes the first two stanzas of *Vandemataram* as the National Song.

3 Though originally written as a standalone lyric, it was published in 1875 as a filler for a blank page in the Bengali journal *Bongodorshon* edited by Chatterji himself.

Vandemataram was inserted into his 1882 novel *Anandamath*. The lyric was primarily disseminated to the British Indian reading public through Chatterji's novel. See Sugata Bose, "Nation as Mother: Representations and Contestations of 'India' in Bengali Literature and Culture," in *Nationalism, Democracy, and Development: State and Politics in India*, eds. Sugata Bose and Ayesha Jalal (New Delhi: Oxford University Press, 1997), 50–75.

4 Surendranath Banerjea, for instance, reminisces about singing *Vandemataram* as well as British proscriptions against it in his memoir. See Surendranath Banerjea, *A Nation in Making: Being the Reminiscences of Fifty Years of Public Life* (Bombay: Oxford University Press, 1925), 186–92. For a book-length study of *Vandemataram*, see S. Bhattacharya, *Vande Mataram: The Biography of a Song* (New Delhi: Penguin Books, 2003).

5 My use of "geo-body" draws on the work of the historian Thongchai Winichakul. See Thongchai Winichakul, *Siam Mapped: A History of the Geo-Body of a Nation* (Honolulu: University of Hawaii Press, 1994). For a delineation of the development of a gendered geo-body of Bharat Mata or Mother India, see Sumathi Ramaswamy, *The Goddess and the Nation: Mapping Mother India* (Durham, NC: Duke University Press, 2010).

6 See chapter 1 for a discussion of *Anandamath*.

7 Sisir Kumar Das, *The Artist in Chains: The Life of Bankimchandra Chatterji* (New Delhi: New Statesman Publishing, 1984), 220.

8 Ram Gopal, *A Political History of Indian Muslims, 1858–1947* (New Delhi: Criterion Publications, 1988), 256–257. Bhattacharya notes that the lyric became a key issue during the Nehru-Jinnah talks of 1938. See Bhattacharya, *Vande Mataram*, 10–18.

9 Bhattacharya, *Vande Mataram*, 21.

10 Sumit Sarkar documents Tagore's growing aversion to the majoritarianism of the Indian freedom struggle in the aftermath of the protests against the 1905 Partition of Bengal. See Sumit Sarkar, *The Swadeshi Movement in Bengal, 1903–1908* (New Delhi: People's Publishing House, 1973).

11 Rabindranath Tagore, "Letter 314. To Subhas Chandra Bose," in *Selected Letters of Rabindranath Tagore*, eds. Krishna Dutta and Andrew Robinson (Cambridge: Cambridge University Press, 1997), 487.

12 *Vandemataram* has been typically understood as a historical document. In addition to Bhattacharya, *Vande Mataram*, see A. G. Noorani, "Vande Mataram: A Historical Lesson," *Economic and Political Weekly* 8, no. 23 (1973): 1039–1043. Scholars have also extensively studied the song's articulation of gender, especially in constructing notions of India and the Indian woman. In addition to Ramaswamy, *The Goddess and the Nation* and Tithi Bhattacharya, "Tracking the Goddess: Religion, Community, and Identity in the Durga Puja Ceremonies of Nineteenth-Century Calcutta," *The Journal of Asian Studies* 66 (2007): 919–962, see Jasodhara Bagchi, "Representing Nationalism: Ideology of Motherhood in Colonial Bengal," *Economic and Political Weekly* 25, no. 42/43 (1990): WS65–WS71; Lise McKean, "Bharat Mata: Mother India and Her Militant Matriots," in *Devī: Goddesses of India*, eds. John Stratton Hawley and Donna Marie Wulff (Berkeley: University of California Press, 1996), 250–280; Tanika Sarkar, "Birth of a Goddess: 'Vande Mataram,' *Anandamath*, and Hindu Nationhood," *Economic and Political Weekly* 41, no. 37 (2006): 3959–3969; André J. P. Elias, " 'Vande Mataram!': Constructions of Gender and Music in Indian Nationalism," *Asian Music* 48, no. 2 (2017): 90–110. Among the few studies that examine *Vandemataram* as a literary text, especially its *mediation* of history and politics is Bose, "Nation as Mother"; J. Lipner, " 'Icon and Mother': An Inquiry into India's National Song," *The Journal of Hindu Studies* 1, no. 1–2 (2008): 26–48; Sudipta Kaviraj, "Making of a Language of Patriotism in Modern Bengali," in *Nationalism and Ethnosymbolism: History, Culture and Ethnicity in the Formation of Nations*, eds. Athena S. Leoussi and Steven Grosby (Edinburgh: Edinburgh University Press, 2007), 248–264; Sudipta Kaviraj,

The Unhappy Consciousness: Bankimchandra Chattopadhyay and the Formation of Nationalist Discourse in India (New Delhi: Oxford University Press, 1995). *Jana Gana Mana*, on the other hand, has not really been the subject of any sustained critical examination. Two notable exceptions are Prabodh Chandra Sen, *India's National Anthem* (Calcutta: Vishva-Bharati, 1972) and Geeta Chattopadhyay, *Bangla Swadeshi Gaan* [বাংলা স্বদেশী গান] (New Delhi: Delhi University, 1983).

13 For his conception of worldliness, see Edward Said, *Culture and Imperialism* (New York: Vintage Books, 1994), 3–61.

14 In my use of the terms "filiation" and "affiliation," I draw on Edward Said's delineation of the terms. See Edward Said, "Secular Criticism," in *The Text, the World and the Critic* (Boston: Harvard University Press, 1984), 1–30.

15 Partha Chatterjee, *The Nation and its Fragments: Colonial and Postcolonial Histories* (Princeton, NJ: Princeton University Press, 1993), 5–6.

16 Nicholas Brown, *Utopian Generations: The Political Horizon of Twentieth-Century Literature* (Princeton, NJ: Princeton University Press, 2005), 10.

17 This is the song from the Hindi film *Upkar* (Dir: Manoj Kumar, 1967) that continues to be popular today.

18 The terms "performative" and "pedagogic" would perhaps recall for the reader Homi Bhabha's formulation of the nation. See Homi K Bhabha, "DissemiNation: Time, Narrative, and the Margins of the Modern Nation," in *Nation and Narration*, ed. Homi K. Bhabha (London: Routledge, 1990), 291–332. My use of these terms, however, is more indebted to Marxist theorizations of ideology that understand it as a structuring ensemble that *constitutes subjects* in addition to *understanding* the world (pedagogic) and making them *act* (performative) in specific ways. For an overview of Marxist understandings of ideology, see Terry Eagleton, *Ideology, An Introduction* (London: Verso, 1991).

19 My use of "allegory" here draws on Walter's Benjamin's theorization of the concept. See Walter Benjamin, *The Origin of German Tragic Drama*, trans. John Osborne (London: Verso, 1998).

20 For the relationship between speech and writing, see Walter J. Ong, *Orality and Literacy: The Technologizing of the Word* (London: Routledge, 2002). Also see Jacques Derrida, *Of Grammatology*, trans. Gayatri Chakravorty Spivak (Baltimore: Johns Hopkins University Press, 2016). For a more global view of the relationship between orality and literature, see Ngũgĩ wa Thiong'o, *Globalectics: Theory and the Politics of Knowing* (New York: Columbia University Press, 2012).

21 See Walter Benjamin, "On the Mimetic Faculty," in *One-Way Street and Other Writings*, trans. Edmund Jephcott and Kingsley Shorter (London: NLB, 1979).

22 Suvir Kaul, *Poems of Nation, Anthems of Empire: English Verse in the Long Eighteenth Century* (Charlottesville: University Press of Virginia, 2000).

23 It bears stressing that literary and cultural texts produced by Indians in the colonial era, across genres and media, were often quite supportive of British colonial rule resulting in a significant corpus of what can only be called *comprador* literature.

24 Kaul, *Poems of Nation*, 1.

25 James Thomson, "Rule Britannia," in *The Oxford Book of Eighteenth Century Verse*, ed. David Nichol Smith (London: Oxford University Press, 1951), 252.

26 Ibid.

27 Kaul, *Poems of Nation*, 2.

28 Denis Cosgrove, "Prospect, Perspective and the Evolution of the Landscape Idea," *Transactions of the Institute of British Geographers* 10, no. 1 (1985): 55. While I am using "landscape" here to signify a spatial view or a "way of looking," this is only one of the diverse ways in which the term has been theorized across disciplines. It has been conceptualized, variously, as the built environment of a place; a habitus; a dwelling; a way of seeing; and a means of spatial representation. Some of the influential theorizations of "landscape" can be found in texts such as: Raymond Williams, *The Country*

and the City (New York: Oxford University Press, 1973); John Barrell, *The Dark Side of the Landscape: The Rural Poor in English Painting, 1730–1840* (Cambridge: Cambridge University Press,1983); Denis Cosgrove, *Social Formation and Symbolic Landscape* (London: Croom Helm, 1984); Kōjin Karatani, *Origins of Modern Japanese Literature* (Durham, NC: Duke University Press, 1993); Don Mitchell, *The Lie of the Land: Migrant Workers and the California Landscape* (Minneapolis, MN: University of Minnesota Press, 1996); W. J. T. Mitchell, "Imperial Landscape," in *Landscape and Power*, ed. W. J. T. Mitchell (Chicago: University of Chicago Press, 2002), 5–30; Tim Ingold, *The Perception of the Environment: Essays on Livelihood, Dwelling and Skill* (London: Routledge, 2002); Kenneth Olwig, *Landscape, Nature, and the Body Politic: From Britain's Renaissance to America's New World* (Madison: University of Wisconsin Press, 2002).

29 Thomson, "Rule Britannia," 253.

30 See Theodore Adorno, *Aesthetic Theory*, eds. Gretel Adorno and Rolf Tiedemann, trans. Robert Hullot-Kentor (London: Continuum, 2002).

31 Reginald Heber, "Before a Collection for the Society for the Propagation of the Gospel," in *The Poetical Works of Reginald Heber* (Boston: Little Brown & Co., 1853), 86–88.

32 For a discussion of these views of wasted plenty in the context of imperialism, see David Spurr, *The Rhetoric of Empire: Colonial Discourse in Journalism, Travel Writing, and Imperial Administration* (Durham, NC: Duke University Press, 1993). Also see Vinay Gidwani, " 'Waste' and the Permanent Settlement in Bengal," *Economic and Political Weekly* 27, no. 4 (1992): PE39–PE46.

33 Thomas Carlyle, "Occasional Discourse on the Negro Question," *Fraser's Magazine for Town and Country* XL, no. CCXL (December 1849): 675. I want to point out that when Carlyle refers to "pumpkins," he means watermelons.

34 Ibid.

35 The ethno-linguistic homeland of Bengal, called *Bongo* [বঙ্গ] or *Bongodesh* [বঙ্গদেশ] in Bengali, comprises the region in South Asia where Bengali, in its diverse dialects, is spoken. At present, it comprises the Republic of Bangladesh, the Indian province of West Bengal, and the Cachar regions of the Indian province of Assam. At the time of *Vandemataram* composition, Bengal signified the areas made up of present-day Bangladesh as well as the Indian provinces of West Bengal, Assam, Bihar, Jharkhand, and Odisha with the exception of Cooch Behar, which were princely states.

36 According to the stanzaic enumeration of *Vandemataram* by Julius Lipner, stanzas 1, 2, 4, 7, 8, 9 are in Sanskrit; stanzas 3 and 5 are in both Sanskrit and Bengali; stanza 6 is entirely in Bengali. The postcolonial Indian nation-state adopted only stanzas 1 and 2 of the original as its National Song. Julius Lipner, *Introduction to Anandamath or The Sacred Brotherhood,* by Bankim Chandra Chatterji, trans. Julius Lipner (New York: Oxford University Press, 2005), 85.

37 I am using Julius Lipner's translation of the song. For the translation, see Lipner, introduction, 84–85.

38 For a global history of famines in the Victorian era, see Mike Davis, *Late Victorian Holocausts: El Niño Famines and the Making of the Third World* (London: Verso, 2002).

39 B. M. Bhatia, *Famines in India; A Study in Some Aspects of the Economic History of India, 1860–1965* (Bombay: Asia Publishing House, 1967), 8. For studies on famines in colonial India see Amartya Sen, *Poverty and Famines: An Essay on Entitlement and Deprivation* (Oxford: Oxford University Press, 1983) and Janam Mukherjee, *Hungry Bengal: War, Famine and the End of Empire* (New York: Oxford University Press, 2015). For the literary registration of natural calamities, including famines, in the Victoria era, see Upamanyu Pablo Mukherjee, *Natural Disasters and Victorian Empire: Famines, Fevers and the Literary Cultures of South Asia* (Basingstoke: Palgrave Macmillan, 2013).

40 Ira Klein, "Death in India, 1871–1921," *The Journal of Asian Studies* 32, no. 4 (1973): 640.

41 Williams, *The Country and the City*, 120. Also see Barrell, *The Dark Side*.

42 For explication of landscape as noun and verb see W. J. T. Mitchell, "Introduction," in *Landscape and Power*, ed. W. J. T. Mitchell (Chicago: University of Chicago Press, 2002), 1–4.

43 Williams, *The Country and the City*, 32.

44 Commenting on the occupations of British Indian subjects, the 1871 census notes: "the largest class is that of persons engaged in agriculture, including those tending or dealing in animals. The number of persons returned under this head is nearly 37½ millions, and forms three-fifths of the entire population." See Henry Waterfield, *Memorandum on the Census of British India 1871–72* (London: Her Majesty's Stationary Office, 1875), 32.

45 Property rights ("rule of property") in land was first introduced by the English in Bengal through the "Permanent Settlement" in 1793. This was followed by the Ryotwari system in the Bombay and Madras Presidencies, and the Mahalwari system in northern India during the 1820s. The introduction of property rights transformed the Indian social formation that witnessed the emergence of new socio-economic classes, especially the largely upper-caste rentier class of colonial Bengal who called themselves the "bhadralok" [ভদ্রলোক], which connotes the sense "refined folk." For the Permanent Settlement, see Ranajit Guha, *A Rule of Property for Bengal: An Essay on the Idea of Permanent Settlement* (Paris: Mouton, 1963). Also see Sirajul Islam, *The Permanent Settlement in Bengal: A Study of Its Operation 1790–1819* (Dhaka: University Press, 1979) and Andrew Sartori *Liberalism in Empire: An Alternative History* (Berkeley: University of California Press, 2014). For discussions on the impact of the *bhadralok* on the cultural sphere of Bengal (and South Asia), see Tithi Bhattacharya, *The Sentinels of Culture: Class, Education, and the Colonial Intellectual in Bengal, 1848–85* (New Delhi: Oxford University Press, 2005); Chatterjee, *Nation and its Fragments*; and Andrew Sartori, *Bengal in Global Concept History: Culturalism in the Age of Capital* (Chicago: University of Chicago Press, 2008).

46 Kaviraj, "Making of a Language," 253.

47 Bose, "Nation as Mother," 53.

48 The 1871 census lists the total population of Bengal as 62,680,633. The figure includes the area under British administration as well as Feudatory States under the jurisdiction of Bengal. See Waterfield, *Memorandum*, 5. According to the 1881 census, the population of Bengal (areas under British rule as well as Princely States) was 69,536,861. See W. Chichele Plowden, *The Indian Empire Census of 1881. Statistics of Population Vol. II* (Calcutta: Superintendent of Government Printing, 1883), 1.

49 Waterfield, *Memorandum*, 16.

50 The 1871 census notes: "The 5 millions of 'Others' are chiefly composed of the hill tribes and aborigines in the Central Provinces, Bengal and Assam, Berar, and British Burma; *but it is very difficult to draw the line Between Hindooism and the rude religion of some of these tribes*, and very possibly many have been classed under the one, when they might with equal propriety have been ranked in the other category." Waterfield, *Memorandum*, 19. Italics mine.

51 See Joya Chaterji, *Bengal Divided: Hindu Communalism and Partition, 1932–1947* (New Delhi: Cambridge University Press, 2006).

52 The word "desh" [দেশ] in Bengali, like the English word "country," retains a sense of semantic plurality that can be, and is, used to signify space in multiple registers: as place in all its rootedness exemplified by the statement "amar *desh* Birbhumey" [আমার দেশ বীরভূমে] that means "I am from Birbhum"; as an abstract territorial expanse evidenced in a compound such as poschim*desh*, [পশ্চিমদেশ] or "the western lands"; and as the modern space of the territorialized nation-state as in the name of South Asia's newest nation-state, Bangla*desh* [বাংলাদেশ].

53 See Bhattacharya, "Tracking the Goddess," 920.

54 The idea of "civil religion" was conceptualized by Jean-Jacques Rousseau who understood it as an alternative to the state-sponsored Catholicism of France. This concept has been particularly influential in the American context and indicates a type of bourgeois civic-mindedness that is an alternative to an overt religious culture. See Jean-Jacques Rousseau, *The Social Contract and Other Later Political Writings, Book IV*, ed. and trans. V. Gourevitch (Cambridge: Cambridge University Press, 1997), 121–152. For engagements with the secularization of Hinduism in the context of British India, see Sandeep Banerjee and Subho Basu, "Secularizing the Sacred, Imagining the Nation-Space: The Himalaya in Bengali Travelogues, 1856–1901," *Modern Asian Studies* 49, no. 3 (2015): 609–649, and Tanika Sarkar, *Hindu Wife, Hindu Nation: Community, Religion, and Cultural Nationalism* (Bloomington: Indiana University Press, 2001), in addition to Bhattacharya, "Tracking the Goddess."

55 See Bhattacharya, "Tracking the Goddess." One of the vignettes from the collection of sketches on Calcutta life from 1862 modelled on Charles Dickens's *Sketches by Boz* called *Hutoom Pyanchar Noksha* [হুতোম প্যাঁচার নকশা] meaning "sketches by Hutoom the Owl" has the "barowaari" [বারোয়ারী] or public Durga festival as its subject. See Kaliprasanna Sinha, *The Observant Owl: Hootum's Vignettes of Nineteenth-Century Calcutta*, trans. Swarup Roy (Ranikhet: Black Kite, 2008).

56 Charles Malamoud, "Bricks and Words: Observations on the Bodies of the Gods in Vedic India," in *Cooking the World: Ritual and Thought in Ancient India*, trans. David White (New Delhi: Oxford University Press, 1996), 209.

57 Ibid.

58 Ibid., 210–212. Also see Hillary Rodrigues, *Ritual Worship of the Great Goddess: The Liturgy of the Durga Puja with Interpretations* (Albany: State University of New York Press, 2003), 168.

59 Kaviraj, "Making of a Language," 253.

60 Lipner, "Icon and Mother," 33.

61 Ibid.

62 Ibid.

63 Rabindranath Tagore, "Bongomata" [বঙ্গমাতা], in *Chaitali* [চৈতালি] (Calcutta: Vishwabharti, 1896), 58. The translation is mine. The original lines are:

> পুন্যে পাপে দুঃখে সুখে পতনে উত্থানে
> মানুষ হইতে দাও তোমার সন্তানে
> হে স্নেহার্ত বঙ্গভূমি, তব গৃহক্রোড়ে
> চিরশিশু করে আর রাখিও না ধরে . . .
> সাতকোটি সন্তানেরে হে মুগ্ধ জননী
> রেখেছ বাঙালি করে মানুষ করনি |

64 Dwijendralal Ray, "Bharotborsho" [ভারতবর্ষ], in *Gaan* [গান] (Calcutta: Bengal Medical Library, 1915), 3. The translation is mine. The original lines are:

> যেদিন সুনীল জলধি হইতে উঠিলে জননী ভারতবর্ষ
> সেদিন বিশ্বে সে কি কলরব, সে কি মা ভক্তি, সে কি মা হর্ষ . . .

65 See Sen, *India's National Anthem*, 23–24.

66 This is my translation of *Jana Gana Mana*. The translation of the entire lyric is given in the Appendix. For the Bengali text of *Jana Gana Mana*, see Rabindranath Tagore, "Jonoganomono-odhinayako" [জনগনমনঅধিনায়ক], in *Gitobitan* [গীতবিতান] (Calcutta: Vishwabharati, 1993), 249–250. For Tagore's own translation of the lyric, see Sen, *India's National Anthem*, 4–5.

67 Chris Bayly, *Origins of Nationality in South Asia: Patriotism and Ethical Government in the Making of Modern India* (New Delhi: Oxford University Press, 1998), 11.

68 For the Bengali text of the poem, see Rabindranath Tagore, "Bharot-tirtho" [ভারততীর্থ], in *Gitobitan* [গীতবিতান] (Calcutta: Vishwabharati, 1993), 251.

69 Rabindranath Tagore, "Pilgrimage to India," in *Rabindranath Tagore: Selected Poems* ed. and trans. Sukanta Chaudhuri (New Delhi: Oxford University Press, 2004), 200–201.

70 Jahan Ramazani, *A Transnational Poetics* (Chicago: Chicago University Press, 2009), 146.

71 Ibid., 202.

72 The mother here refers not to a Hindu deity but the image of space in a non-religious anthropomorphic vein, as in Tagore's song to Bengal as a mother – "amar shonar Bangla" [আমার সোনার বাংলা], typically translated as "My Golden Bengal" or "My Bengal of Gold" – that would be adopted as the national anthem of Bangladesh. See Rabindranath Tagore, "amar shonar Bangla" [আমার সোনার বাংলা] *Gitobitan* [গীতবিতান] (Calcutta: Vishwabharati, 1993), 243. For an English translation of the song, see Rabindranath Tagore, "My Bengal of Gold," in *The Essential Tagore*, eds. Fakhrul Alam and Radha Chakravarty, trans. Fakhrul Alam (Cambridge: The Belknap Press of Harvard University Press, 2011), 330–331.

73 This can be seen in the Bengali renditions of *The Ramayan* and *The Mahabharat* where the translators, namely, Krittibas Ojha and Kashiram Das refer to themselves by name in the concluding line of each section. See for instance, Krittibas Ojha, *Shochitro Shoptokando Ramayon* [সচিত্র সপ্তকান্ড রামায়ন], ed. Ramananda Chatterjee (Calcutta: Probashi Press, 1926).

74 This device is called *ankita* in the Kannada *vachanas* (lyrics in free verse) to Shiva by the Veerasaiva poet-saints who address their compositions to Shiva. Shiva is addressed by a specific name or phrase, which in turn identifies the poet. For instance, Basava calls Shiva *kudalasangamadeva* ["Lord of the Meeting Rivers"] or *guhesvara* ["Lord of Caves"] in his *vachanas*. For a discussion of the *ankita* in Lingayat poetry, see K. V. Zvelebil, *Introduction to The Lord of the Meeting Rivers: Devotional Poems of Basavanna* (New Delhi: Motilal Banarsidass, 1984), 1–4.

75 For details of the controversy, see Sen, *India's National Anthem*, 6–22.

76 Quoted in Sen, *India's National Anthem*, 7.

77 Ibid.

78 While the set of practices called Hinduism is typically understood to be of a cataphatic nature, engaged in concrete representations of the divine and idol worship, there also exists a strong apophatic dimension to it. For instance, in the dialogue between King Janaka and the sage Yajnavalkya in the *Brhad-aranyaka Upanishad* (IV.2.4) the sage speaks of the self ("atman") through typical registers of apophatic or negative theology. Yajnavalkya says, "the self is not this, not this. He is undestructible [sic] for he cannot be destroyed. He is unattached for he does not attach himself. He is unfettered, he does not suffer, he is not injured." See *Brhad-aranyaka Upanishad* in *The Principal Upanishads*, ed. and trans. S. Radhakrishnan (London: George Allen & Unwin Ltd., 1953), 253–254.

79 My use of "heteroglossia" (sometimes translated as "polyglossia") derives from the delineation of the term by Mikhail Bakhtin. See M. M. Bakhtin, "Discourse in the Novel," in *The Dialogic Imagination: Four Essays by M. M. Bakhtin*, ed. Michael Holquist, trans. Caryl Emerson and Michael Holquist (Austin: University of Austin Press, 1981), 259–422.

80 See Chatterjee, *The Nation and its Fragments*.

81 Dipesh Chakrabarty, for instance, has a rather reductive understanding of the "spiritual" that for him refers unambiguously, and rather simplistically, to gods and spirits. He writes, "I take gods and spirits to be existentially coeval with the human and think from the assumption that the question of being human involves the question of being with gods and spirits." See Dipesh Chakrabarty, *Provincializing Europe: Postcolonial Thought and Historical Difference* (Princeton, NJ: Princeton University Press, 2000), 16.

82 Chatterjee's insistence of the separation of the material and spiritual domains in Asia and Africa for the project of fashioning their nationalisms appears to ignore the ways in which the "spiritual" have been deployed to make claims to social justice – and a more equitable material life – across the globe, including in the "west." In Latin America,

this is perhaps most extensively engaged in what has come to be called liberation theology. See, for instance, Gustavo Gutiérrez, *A Theology of Liberation: History, Politics, and Salvation* (Maryknoll: Orbis Books, 1973). Messianism and millenarianism structure the Marxist theorizations of Walter Benjamin and Ernst Bloch, especially the latter's formulation of the "warm stream of Marxism." See Bloch, *The Principle of Hope, Vol. 1*, 209. In the context of Britain, literary critics and historians have examined how poets such as John Milton and William Blake draw on radical visions, and energies, of a variety of Christian sects, namely, the Ranters, Diggers, Quakers, Behmenites, Muggletonians, and Southcottians who deploy the Christian spiritual universe and vocabulary to imagine a more just social world. A classic study on John Milton in this context is Christopher Hill, *Milton and the English Revolution* (New York: Viking Press, 1978). For analogous engagements with William Blake, see A. L. Morton, *The Everlasting Gospel: A Study in the Sources of William Blake* (New York: Haskell House, 1966); E. P. Thompson, *Witness Against the Beast: William Blake and the Moral Law* (New York: New Press, 1993); and Jon Mee, *Dangerous Enthusiasm: William Blake and the Culture of Radicalism in the 1790s* (Oxford: Clarendon Press, 1994). For an account of the radical Christian sects active around the English Civil War, see Christopher Hill, *The World Turned Upside Down: Radical Ideas During the English Revolution* (Harmondsworth: Penguin Books, 1975). For a sense of the interplay between the spiritual and material in the writings of Ranter preachers from seventeenth-century England such as Abiezer Coppe and Jacob Bauthumley, see the excellent collection: Nigel Smith (ed.), *A Collection of Ranter Writings from the 17th century* (London: Junction Books, 1983).

83 This underscores the charge of orientalism against theorists such as Partha Chatterjee and Dipesh Chakrabarty by critics such as Vasant Kaiwar and Vivek Chibber. For a robust critique of Chatterjee (and Chakrabarty), see Vasant Kaiwar, *The Postcolonial Orient: The Politics of Difference and the Project of Provincialising Europe* (Leiden: Brill, 2014). Another critique, if less successful on the question of the subalternists' auto-orientalism, is Vivek Chibber, *Postcolonial Theory and the Specter of Capital* (London: Verso, 2012).

84 William Blake, *Milton: A Poem in 2 Books*, in *The Complete Poetry and Prose of William Blake*, ed. William V. Erdman, Newly Revised Edition (New York: Anchor Books, 1988), 95–96.

4 A sense of place

Narrating knowable communities

Novels posit imaginations of communities that are located – or better still, *grounded* – in their imagined geographies. Raymond Williams suggests that most novels are "in some sense knowable communities" where "the novelist offers to show people and their relationships in essentially knowable and communicable ways."[1] In positing communities as knowable and known, they present socio-spatial formations as contingent wholes that gesture to the larger space of the colony or postcolony. And in presenting the post/colony as a spatial community, they inscribe those spaces with a lived dimension and provide readers, in effect, with *a sense of place*. In the case of British India, novels *narrate* the socio-spatial configuration that make up that territorial expanse in either a colonial or a counter-colonial vein, drawing out social relations and subjectivities – *lives*, in sum – as they are lived in that space. In this chapter I examine how three novels from British India – Rudyard Kipling's *Kim* (first published in 1901), Rabindranath Tagore's *Gora* (first published in 1909), and Rokeya Sakhawat Hossain's *Padmarag* (first published in 1924) – narrate the colonial space as a socio-spatial entity in a utopian vein paying particular attention to their depiction of life and subjectivities as lived.

Kim was first published serially from December 1900 in *McClure's* magazine. It appeared in *McClure's* as well as *Cassell's* magazine throughout 1901 before being published as a book in October 1901. Kipling presents the British Indian colony as a spatial contradiction: as both a territorial expanse ruled by a colonial state, that is, a functional space that is to be governed, as well as a lived space or a dwelling. This sets out insider-outsider perspectives on space and community simultaneously that are negotiated in the novel through the figure of the Irish foundling protagonist Kim. *Gora* appeared serially in 1907 in the pages of the Bengali literary magazine *Probashi*, before being published as a book in 1909. Translated by W. W. Pearson, its English edition was published from London in 1910. Tagore's novel also has an Irish foundling as its protagonist. But if Kipling's *Kim* is both an insider and an outsider, Tagore's Gora is an outsider made insider: a "naturalized" Indian who, over the course of the novel, learns that to be "Indian" is not necessarily to be an orthodox Hindu. Readers are made aware of Gora's Irish provenance near the opening of the narrative. While this necessarily ironizes Gora's desire to be an orthodox Hindu, it also constructs an ideal of Indian space and community as open and welcoming that makes no distinction between the "other" and the "self." This for Tagore, however, is an ideal – the utopian horizon, if you will – that has to be struggled for and actualized and not the essential or normative characteristic of the socio-spatial entity that is India. This is made evident in the novel by showing Indian space and society fissured in terms of gender, class, caste, religion, and sect. His India ranges from within the home to the world outside, incorporating views of the city and the country that are populated by the Indian elite – anti-colonial and comprador – and the British. Tagore makes visible Muslims, peasants, untouchables, and the underclass, repeatedly marking India as a fundamentally plural entity: a spatialized heteroglossia. The socio-spatial entity of India thus becomes a site where Hindu parochialism, universal humanism, anti-colonial nationalism, casteism, sexism, and the struggle for women's emancipation co-exist. Hossain's text, while focused on Bengal, in also invested in drawing out the plurality of the socio-spatial formation. In addition, it brings an explicitly feminist perspective on to the colonial space and strives to reconceptualize it beyond the confines of patriarchy. As Barnita Bagchi notes in the introduction to her translation of *Padmarag*, the novella presents a complex educational and philanthropic female utopia where Hossain's vision "reveals that women, be they Hindus, Brahmos, Muslims or Christians, black or white, are all victims of patriarchal oppression."[2] It narrates the development of the eponymously named female protagonist Padmarag (also called Siddika and Zainab), and signals community through the space of Tarini Bhavan, a shelter for destitute women that is wholly run and administered by women. Tarini Bhavan allegorizes the utopian possibility of an egalitarian post-colony where women are at par with men and control their own lives.

My constellation of these texts also throws up distinct views on space and community that are marked by colonial and counter-colonial spatial desires, and points of resonance between these authors who were also contemporaries. Kipling and Tagore were unquestionably the two most celebrated authors on either side of the colonial divide, while Hossain was a pioneering feminist and advocate of women's issues in British India in addition to being a writer in her own right.[3] Typically known in the Euro-American academy as the author of the utopian short story "Sultana's Dream," Hossain's oeuvre nevertheless remains largely ignored by scholars of postcolonial literature and theory.[4] Kipling was awarded the Nobel Prize in Literature in 1907, becoming the first English-language writer to be so recognized. Tagore was awarded the literature Nobel in 1913, becoming the first Asian (and non-white) person to be given the prize. If Kipling was the "bard of Empire," then Tagore was, in the words of Mohandas Gandhi, "a sentinel warning . . . against the approach of enemies called Bigotry, Lethargy, Intolerance, Ignorance, Inertia and other members of that brood."[5] And while Tagore wrote extensively about education, and set up a school (Patha Bhavan), university (Vishwabharati), and a vocational training institute (Sriniketan), Hossain set up a school for girls in Bhagalpur in 1909 and moved it to Calcutta the following year. The school continues to thrive today as the Sakhawat Memorial Girls' High School and is "the most enduring testimony to Rokeya's vision and competence as an educationist."[6]

Spatializing contradiction: Kipling's *Kim*

Kim narrates the travels and adventures of its eponymous Irish foundling protagonist Kimball O'Hara, or Kim, as he is known to the world, who is introduced to readers as a street urchin growing up in the streets of Lahore and living off the kindness of strangers. At the start of the novel, he meets Teshoo Lama outside the Lahore Museum. The Lama has come to India from Tibet in search of the river that emerged where the Buddha's arrow fell on earth, and Kim decides to travel with him to help in his quest. In the course of his peregrinations, he delivers a message to the English for his friend Mahbub Ali, unwittingly helping the colonial state in the Great Game.[7] He subsequently meets "The Mavericks," his deceased father's regiment, and its officers send Kim to St. Xavier's School in Lucknow: to turn him into a "sahib" and train him to work for the colonial Indian administration. In the course of the narrative, Kim comes in contact with Colonel Creighton of the Great Trigonometric Survey of India, and other members of the British Indian colonial establishment such as Mr. Lurgan (Lurgan Sahib) and Huree Chunder Mookerjee (Huree Babu). Kim eventually plays an active role in the Great Game, assisting Hurree Babu in foiling the espionage activities of French and Russian spies. At the novel's close, Kim falls ill and is restored to health while the Lama finds his river.

Said finds *Kim* "as unique in the author's life and career as it is in English literature."[8] The novel is indeed unique: it appeared more than a decade after Kipling had left India, and is the author's "only successfully sustained and mature piece of long fiction."[9] The novel is, moreover, "Kipling's celebrated portrait of India," in which the socio-spatial configurations of the British Indian colony are drawn out with care and a touch of nostalgia.[10] Throughout the novel, the space of British India is presented as both the disenchanted state-space of the colony, as well as a lived space, a dwelling. The contradiction that emerges through this dual perspective is central to the novel's socio-spatial imaginary. In the narrative, this contradiction is anchored in the figure of Kim who repeatedly provides readers with an outsider's as well as an insider's perspective on British Indian space and community. The novel opens outside the museum in Lahore, the principal city of the British Indian province of Punjab. Kim is introduced through these lines: "He sat, in defiance of municipal orders, astride the gun Zam-Zammah on her brick platform opposite the old Ajaib-Gher – the Wonder House, as the natives call the Lahore Museum."[11] The lived sense of urban space is underlined in the novel through the figure of Kim – the friend of the world – and his interactions with people around him:

> As he drummed his heels against Zam-Zammah he turned now and again from his king-of-the-castle game with little Chota Lal and Abdullah the sweetmeat-seller's son, to make a rude remark to the native policeman on guard over rows of shoes at the Museum door. The big Punjabi grinned tolerantly: he knew Kim of old. So did the water-carrier, sluicing water on the dry road from his goat-skin bag. So did Jawahir Singh, the Museum carpenter,

> bent over new backing cases. So did everybody in sight except the peasants in the country.[12]

This description shows life as it is lived around the Lahore Museum with Kim at its centre. He is portrayed as someone whom people know and like, as in the case of the native policeman who grins at Kim's "rude remark." The figure of Kim, then, humanizes the space of the city of Lahore. In addition, while readers are given a sense of the city through its built environment, the narrator also presents it as a lived space. The narrator moves from the Museum to the Masonic Lodge ("the big blue and white Jadoo-Gher – the Magic House, as we name the Masonic Lodge") to the Punjab High Court.[13] It is important to note that the buildings are presented not just as edifices, but through the perspective of the locals through the use of terms such as "Ajaib-Gher – the Wonder House" and "Jadoo-Gher – the Magic House." The Punjab High Court is depicted differently, for it is indicated along with "Nila Ram's timber-yard" where "the fragrant the older logs lie seasoning after they have driven down the Ravee."[14] Notice how this description weaves together the visual and olfactory registers. The focalizations on space in this passage occur at two registers: a distanced view of space (signalled through the naming of the buildings) is conjoined with a more experiential and lived sense of place (signalled through names given by the locals, or through the sense of smell). The text does not simply provide two perspectives on space but sets them up in a relation of simultaneity, resulting in a kind of perspectival tension. This perspectival tension, which is a hallmark of the novel and part of its unique socio-spatial perspective, maps on to the tension between the spatial identities of British India as a functional *and* a lived space, that is, to the idea of British India as a *spatial contradiction.*

This sense of spatial contradiction is also evident in the way the novel spatializes time, construing towards British India as anachronistic space. The novel begins, as I have noted, in front of the Lahore Museum. As Kim enters the museum through the turnstile, he notes

> In the entrance-hall stood the larger figures of the Greco-Buddhist sculptures done . . . how long since, by forgotten workmen whose hands were feeling, and not unskilfully, for the mysteriously transmitted Grecian touch. There were hundreds of pieces, friezes of figures in relief, fragments of statues and slabs crowded with figures that encrusted the brick walls of the Buddhist *stupas* and *viharas* . . . and now, dug up and labelled, made the pride of the Museum.[15]

This description of the museum presents it, unsurprisingly, as a space that showcases the past in the present. The past that is marked by the work of "forgotten workmen" now exists "dug up and labelled" as the "pride of the museum." Thus, the museum is presented as a site that anchors the *simultaneity of the non-simultaneous* (the past and the present), marking it as a contradictory space.[16] This also makes the museum an anachronistic space, or a spatial anachronism, and the text appears to hint at an analogy between this local space of the museum and the larger space of the British Indian colony.[17] The temporalization of space is further emphasized as Kim travels with the Lama to the train station and beyond, with the text making repeated assertions about Indians having, and keeping, Oriental time. When Kim receives his message for Colonel Creighton from Mahbub Ali, readers are told that "even an Oriental, with an Oriental's views of the value of time, could see that the sooner it was in the proper hands the better."[18] Likewise, while describing Indian passengers sleeping in the Lahore railway station, readers are told that all "hours of the twenty-four are alike to Orientals, and the passenger traffic is regulated accordingly."[19] Similarly, after Kim had pre-empted an assassination attempt on Mahbub Ali, the latter tells him that he and Kim should proceed for Simla *swiftly*. This is followed up by a sentence imbued with a significant dose of irony: "Swiftly, – as Orientals understand speed, – with long explanations, with abuse and windy talk, carelessly, amid a hundred checks for little things forgotten."[20] This line, as with the other instances in the novel that stress "oriental" time, implicitly point to a normal time of the "occident," or of colonial capitalist modernity.[21] It is of course crucial to note the spatialization of temporal difference here that sets up the spaces of the metropole and the colony as distinct, even as it denies the coevalness of these spaces.[22] But the denial of coevalness is precisely that – a *denial*, and Indians both partake in the temporal registers of colonial capitalist modernity, and they do not. Despite his comments about the Orientals' differential sense of time, the narrative notes that the "3:25 a.m. south bound [train] roared in." This shows that the normative time of capitalist modernity ("3:25 am") had indeed put down roots in the colony. The text seems to gesture to the colony as the spatialization of both normative and aberrant time and thus, a spatial contradiction.

As Kim and the Lama enter the crowded train to make their way from Lahore to Ambala (Umballa), readers are told that their third-class carriage is populated by a "burly Sikh artisan," a blue-turbaned "Hindu Jat," a fat "Hindu money lender," a young "Dogra (Kashmiri) soldier," and a "courtesan" from Amritsar.[23] The train compartment allows the text to articulate a sense of British Indian community where the carriage appears to be a microcosmic representation of British India. Crucially, none of these characters are given names, but are instead represented along the axes of caste, religion, community, and profession, that is, as native types. We have already encountered this style of documentation earlier, in Kipling's depiction of people around the Calcutta port. Here, and unlike in the case of Calcutta, Kipling is documenting Indians and not people from all over the world. In depicting Indians as native types, he shows his reliance on colonial ethnography, most significantly expressed in the collection, *The People of India*. Published between 1868 and 1875, this collection of 468 albumen prints of "native types" owes its origins to Charles Canning and his wife's desire to carry back home a collection of photographic illustrations that depicted "the peculiarities of Indian life."[24] For this, the Governor-General encouraged civilians and army officers to travel with cameras and deposit copies of their photographs with his office. After the 1857 Uprising, the project was given official status and in

1863, the government decided to turn it into a publication documenting Indian communities from across the country. The compendium that emerged, according to Christopher Pinney, displayed "a concern with political loyalty (or its lack) and an ongoing desire to provide practical clues to the identification of groups which had so recently had the opportunity to demonstrate either their fierce hatred of British rule or their acquiescence."[25] The language of the preface is particularly illuminating:

> The great convulsion of 1857–58. . . imparted a newer interest to the country which had been the scene, and to the people who had been the actors in these remarkable events. When, therefore, the pacification of India had been accomplished, the officers of the Indian services, who had made themselves acquainted with the principles of practices of photography . . . *went forth and traversed the land in search of interesting subjects*.[26]

The phrase "pacification of India" obscures the true import of that act and screens the inherent violence that marked the quelling of the 1857 Uprising. Moreover, the use of words and phrases such as "actors," "scene," "remarkable events," and "interesting subjects" sanitizes history and collective memory, reconstituting India and Indian history from a British viewpoint as objects of curiosity, colonial study, and knowledge production. *The People of India* puts the featured "native types" on virtual display through the photographs, as if being exhibited in a museum. In addition, these "native types" are labelled erratically according to categories of caste, tribe, or sect, with the captions supplemented with more detailed descriptions that narrate the characteristics of these Indians. Interestingly enough, one of the "types" in the first volume of this collection is a Buddhist Lama. But even as Kipling deploys the ethnographic modality to describe the people in the train compartment, it is nevertheless inscribed as a lived space. Its occupants are depicted as humane, laughing, screaming, and speaking loudly. And once again we find in the depiction of spatial community in *Kim* a tension between a view that is distant and aloof, and one that is more situated and involved.

The perspectival tension continues when Kim travels with the Lama through the Grand Trunk Road, the most important highway in northern and eastern South Asia. As they approach the road, an old soldier, who has been their companion to the road, exclaims, "See . . . the Great Road which is the backbone of all Hind [India] . . . In the days before the rail-carriages the Sahibs travelled up and down here in hundreds. Now there are only country-carts and such like."[27] The old soldier then proceeds to catalogue the Indians, quite like his colonial masters. He says, "Look! Brahmins and chumars [tanners of low caste], bankers and tinkers, barbers and bunnias [traders], pilgrims and potters – all the world going and coming. It is to me a river."[28] It is illuminating that the soldier describes Indians along the axes of caste and occupation, displaying the native's internalization of the colonial, ethnographic modality of enumeration. As Kim walks down the Grand Trunk Road with the Lama, he too is soaked in the atmosphere of seeing all of India on the move. It is for him a "broad, smiling river of life" that offers "new people and new sights at every stride – castes he knew and castes that were altogether out of his experience."[29] But even as he calls the road a "river of life," this vision of community is profoundly mediated through the lens of colonial ethnography. Kim notes several groups of people, all of which implicitly allude to characters from *The People of India*. The first group of people he meets – "a troop of long haired, strong scented Sansis [nomadic people] with baskets of lizards and other unclean food on their backs" – are depicted in not-so-glowing terms in the collection.[30] Writing about the Sansis, *The People of India* notes that "among them are found professors of normal hereditary crime in every possible shape, from the remorseless Thug and Dacoit to the cheats, pickpockets, and petty thieves who follow their calling in every bazar or fair of the country."[31] Kim also encounters an Akali [member of a Sikh sect] who is described as a "wild eyed, wild haired Sikh devotee in the blue-checked clothes of his faith."[32] He notes that he was careful "not to irritate that man; for the Akali's temper is short and his arm quick."[33] And here, once again, we see a tension between the simultaneous deployment of two perspectives for depicting the Indian community. This tension not only posits British Indian space as contradictory, but also presents Kim as both an *insider* and an *outsider* to this space, both a part of the community of the British Indians, and not.

This tension between the insider and outsider is most evident in the novel's depiction of space through the landscape form. To *landscape* a space, as I have noted earlier, is to present it as a spatial view; it implies "separation and observation."[34] Throughout the novel, Kim is described as engaging with the space of India as a vast landscape. Travelling down the Grand Trunk Road, for instance, he notes that he was:

> seeing all India spread out to left and right. It was beautiful to behold the many-yoked grain and cotton wagons crawling over the country roads: one could hear their axles, complaining a mile away, coming nearer, to its shouts and yells and bad words they climbed up the steep incline and plunged on to the hard main road, carter reviling carter. It was equally beautiful to watch the people, little clumps of red and blue and pink and white and saffron, turning aside to go to their own villages, dispersing and growing small by twos and threes across the level plain.[35]

In the extract above, the landscape form is deployed right at the outset. Kim is shown as seeing all of India spread out left and right. The ability to see "all India" in one instant points to the cartographic organization of space, and a way of seeing that is predicated on distance and separation. This is followed up with another register of visuality, the view of the "many-yoked grain and cotton wagons crawling over the country roads." But this view from a distance is disrupted with the phrase "coming nearer." The focalization, as it were, zooms in on the carter to tell us of the "shouts and yells and bad words" thereby marking the visual register with the aural. In so doing, it marks the perspective of the *faraway* with that of the *near*. The focalization then pulls out to speak of the British Indian colony and its inhabitants as "little clumps of red and blue and pink and white and saffron." The space, rendered a sea of colour where its inhabitants are turned into specks, is now returned to the landscape perspective. It is turned into a pleasant prospect and Kim, quite like the colonial administrators, is a detached and alienated observer.

The structuring of Indian space as a prospect, to be apprehended through the faculty of vision also meets the cartographic imagination during Kim's travels in the train. As the train passes through the north Indian countryside, Kim observes that "golden, rose, saffron, and pink, the morning mists smoked away across the flat green levels. *All the rich Punjab* lay out in the splendor of the keen sun."[36] Kim grasps the landscape of Punjab as a unity, quite like the conjectured monocular viewer of a landscape. Indeed, his visual appreciation of the landscape of Punjab appears as an expression of the cartographic imagination for it is only possible to see "all the . . . Punjab" at once on a map. But once again, this view is destabilized during his journey through northern India. Kim speaks of strolling through "the hard-worked soil [that] gives three and even four crops a year – through patches of sugar-cane, tobacco, long white radishes and nol khol [kohlrabi] . . . rousing village dogs and sleeping villages at noonday."[37] The insertion of the "hard-worked soil" and the listing of the agricultural produce undermine the space as a prospect, marking it instead as a space of labour and life. Furthermore, owing to the invocation of "village dogs" and "sleeping villages," the space is rendered a communitarian space and a dwelling. This sense of simultaneously being inside and outside the landscape is further accentuated when Kim and the Lama are described as following "the rutted and worn country road that wound across the flat between the great dark-green mango-groves, the line of snow-capped Himalayas faint to the eastward."[38] This line appears to present a normative landscape view that is predicated on a sense of distance and alienation, and structured primarily through the faculty of vision. However, the next line disrupts the stability of this construction as we are told that all of "India was at work in the fields, to the creaking of well-wheels, the shouting of ploughmen behind their cattle, and the clamour of the crows."[39] The stability of the visual register of the landscape form is broken by inserting the aural into the description, and the "creaking," "shouting," and the "clamour" undermines the authority of the visual register of the preceding lines. The subsequent line brings the focus to bear on Kim himself, as we are told that the "pony felt the good influence and almost broke into a trot as Kim laid a hand on the stirrup-leather."[40] This sentence not only highlights Kim's presence, but also locates him squarely within the space, by inserting a tactile and experiential element into the description.

Kim's position in the colonial Indian social formation is unique, for he is both an insider and an outsider to the space of British India. He does not need to "go native," since he can *be* native. At various points in the narrative, a Hindu, a Muslim, and a Buddhist; he can become a member of the high castes or the low; and he can also become "white" when the need arises. Angus Wilson, commenting on Kim, notes that he "unites the knowingness, the cunning, the humour and the appeal of the Dodger, the gentleness and goodness of Oliver Twist, a seemingly impossible task."[41] Kim's fluid subjectivity is also a function of his Irish identity.

The text flip-flops on his racial character throughout, and he is called *white* and *Irish* at different moments in the narrative, suggesting it cannot make up its mind about placing him in relation to the colonizer-colonized divide.[42] After introducing him as "English" in the first page of the novel, he is labelled "Irish" a few pages later. He is reported as saying, "I hate all snakes" with the narrator following up the comments that no "native training can quench the *white* man's horror of the Serpent."[43] When Kim beats the Russian for hitting the Lama, the text notes that the Russian's blow to the Lama had "waked every unknown *Irish* devil in the boy's blood"[44] and finally, when Kim goes through the plans after thwarting the Russian and the Frenchman, the text notes that the "humour of the situation tickled the Irish and the Oriental in his soul."[45] And owing to his ability to be both native and foreigner, Kim collapses the racial – and by extension, the spatial – division between the ruler and the ruled. This becomes a source of tremendous colonial anxiety, and it is hardly surprising that the text spends a substantial amount of energy to draw out his "education" and a project for his "sahibization."[46]

Kim's status of simultaneously being an insider and an outsider structures his lived relationship with the British Indian colony. This is particularly evident when Kim is at St. Xavier's in Lucknow for his education and "sahibization." He notes, for instance, that his companion, the drummer-boy, "styled all the natives 'niggers'; yet servants and sweepers called him abominable names to his face, and, misled by their deferential attitude, he never understood."[47] This points to the limit of the colonizer's ability to know the natives. Kim's ability to transcend this limitation is underscored most poignantly through two markers of native life that were typically baulked at by Europeans, being barefoot and spicy food. At St. Xavier's, Kipling writes, Kim "yearned for the caress of soft mud squishing up between the toes, as his mouth watered for mutton stewed with butter and cabbages, for rice speckled with strong-scented cardamoms, for the saffron-tinted rice, garlic and onions, and the forbidden greasy sweetmeats of the bazars. They would feed him raw beef on a platter at the barrack-school, and he must smoke by stealth."[48] Kim is shown as someone who desires to run about barefoot in the mud; he can stomach the bland beef of the barrack-school but longs for the "mutton stewed with butter and cabbages." And here again we have folded into the figure of Kim, a duality of perspective.

Kim's unique position in the colonial Indian formation also structures his engagement with the people who inhabit it. Kim is shown interacting with natives confidently and naturally, and in the course of the narrative learns how to engage representatives of the colonial state. He is friendly to the Pathan horse dealer Mahbub Ali, and grows in warmth towards the Bengali *babu*, colonial ethnologist, and British Indian spy Huree Chunder Mookerjee. Ali and Mookerjee are both central to the narrative's progress and in Kim's interactions with them, there is more than a hint of colonial stereotyping. Ali is depicted as a Pathan, and while Kipling relies on the colonial notion of the Pathan as a "martial race," there is, however, a genuine affection that shines through in his depiction of their mutual relationship.[49] Likewise, Huree Mookherjee is presented as a "fearful Bengali" but is also an accomplished spy and ethnologist. Not only does he help to train

Kim in his quest for a position in the Great Game, Mookerjee's desire to have an "F. R. S." next to his name, that is, to be a Fellow of the Royal Society is met with approbation from Colonel Creighton, depicted as the most powerful person in the narrative universe of *Kim*. On hearing Mookerjee's ambition, Kipling tells us that Creighton "smiled, and thought the better of Huree Babu, moved by like desire [as himself]."[50] Here too emerges a sympathetic, and almost affectionate, picture of the English-educated Bengali, typically deemed cunning and lazy by the British, an image that Kipling himself helped disseminate in his earlier writings. And towards the end of the narrative when Kim tells Huree Babu that it "were better in Mahbub Ali's hands than a Bengali's," Kipling quite surprisingly gives the Bengali the last word with Huree Babu saying "There are more ways to getting to a sweetheart than butting down a wall."[51]

Ali and Mookerjee who stand in for the Pathans and the Bengalis are, significantly, depicted as loyal native subjects. This is unlike the historical reality of the Pathans and the Bengalis, both social groups who resisted British rule continually, and often violently: the Pathans for much of the nineteenth century, and the Bengalis through the late-nineteenth and the early-twentieth centuries.[52] The spatial desire of Kipling's text appears to reconstitute the recalcitrant native types into the docile subjects of the colonial state and presents the British Indian colony as a community of compradors. But the attempt to reconstitute resistant subjects as compradors is not without its slippages. In describing Huree Babu's encounter with the foreign spies, Kipling writes:

> He became thickly treasonous, and spoke in terms of sweeping indecency of a Government which it forced upon him a white man's education and neglected to supply him with a white man's salary. He babbled tales of oppression and wrong till the tears ran down his cheeks for the miseries of his land.[53]

Kipling's description attempts to ironize the Indian native's complaints against the colonizers. But even though he uses words like "babbled" to undermine the native perspective, the extract above nevertheless highlights, despite itself, a counter-colonial spatial desire. Huree Babu's protest of the "indecency" of the colonial state – of granting him a white man's education without a white man's salary – was indeed an accurate description of the life of native elites under colonialism. Kipling's use of phrases such as "thickly treasonous" and "oppression and wrong," then, appear to be a tacit acknowledgement of counter-colonial spatial desire for decolonization at the exact instance where it is trying to underline colonial hegemony.

The text's construction of British India as a community of compradors is linked to the larger colonial project of maintaining hegemonic control over Indian space, a project into which Kim is inducted. The need to maintain control over India took a renewed shape during the Great Game between Britain and Russia, which forms the backdrop of the novel. Following rumours that Russia might invade British India, the colonial state sent out missions for spying, surveying, and developing geographical measures of Central Asia and especially Tibet. In the late-nineteenth century, the Indian Survey Department under Captain Thomas Montgomerie trained a number of Indians in surveying techniques and sent them on spying missions into Tibet to collect geographical and ethnographic information.[54] These British spies dressed up as Buddhist pilgrims and carried sextants and thermometers in their prayer wheels. Further, they counted their paces, and had rosaries of a hundred beads (instead of the traditional 108) to calculate the distances they travelled. Notable among them were Pandit Nain Singh Rawat, Pandit Krishna, Ugyen Gyatso, and Sarat Chandra Das. Rawat travelled into Tibet in 1865 and reached Lhasa in January 1866, staying there until April 1867. He determined the location and altitude of Lhasa for the first time besides mapping a large section of the River Tsangpo (Brahmaputra). Rawat visited Lhasa again in 1874 and was awarded the Patron's Medal by the Royal Geographical Society in 1877 for his achievements. Krishna reached Lhasa in 1878 and travelled around Tibet before returning to India in 1882. Gyatso and Das journeyed into Tibet for six months in 1879. Das travelled to Tibet again, in 1881, when he visited the Dalai Lama in Lhasa. The British Indian government published Das's reports of his journeys but kept them strictly confidential until 1890. They were subsequently edited and published by the Royal Geographical Society in 1902. Gyatso visited Tibet and Lhasa again in 1883, before returning to India the same year. The maps created from the information the spies collected were later used by the British during their military expedition into Tibet under Francis Younghusband in 1903–1904.[55]

The politics of the Great Game suffuses *Kim*, and Kipling is credited with placing the term "Great Game" firmly in the British imaginary.[56] Published in 1901, the novel appears between the two publications of Das's account of his activities in Tibet. The covert operations in Tibet are most directly mentioned by Creighton when he journeys to Lucknow with Kim. On the train he asks Kim to learn how to make pictures of roads and rivers; he also claims that someday he might ask Kim to "Go across those hills and see what lies beyond" and that there are "bad people living across those hills who will slay the chain-man [surveyor] if he be seen to look like a Sahib."[57] Kim's introduction as a trainee in the activities of the Great Game is supplemented by the reader's knowledge of his great ingenuity with disguises. He could be a Muslim one moment and a Hindu the next; he can also be attired in Buddhist garb and travel as the Lama's *chela* (disciple). The idea of disguise is particularly crucial as it gestures towards British India, much like the preface to the *People of India*, as a theatrical stage. More importantly, it re-inscribes the game ethic in British Indian life. Not only are the diplomatic maneuvers referred to as a game that one plays until one's death, Mahbub Ali also repeatedly refers to Kim in conversations with Creighton as the "young colt" who would one day play polo, and sometimes as a "polo-pony."[58] As Kim graduates to being a player in the game to consolidate British hegemony in South Asia, he transforms from a nomadic entity to one of "cartographic stability," brought into the service of the colonial state through the figure of Creighton.[59]

Despite Kim's transition to cartographic stability, he retains a nomadic entity, that is, his ability to turn native. This results in the narrative's refusal to allow him sexual relations with Indian women. This is evident in his interaction with the "hillwoman" who is revealed to be Lispeth. By naming this female character Lispeth, Kipling makes an inter-textual reference to one of his own short stories.[60] The character from the short story is a concubine of a colonial official who goes away, promising to marry her on his return. Unsurprisingly, the Englishman never returns and this forms the substance of Kipling's story. In *Kim*, Lisbeth narrates this incident to Kim saying:

> Once, long ago, I wore European clothes at the Mission-house yonder . . . Once, long ago. I was *Ker-lis-ti-an* [Christian] and spoke English – as the Sahibs speak it. Yes. My Sahib said he would return and wed me – yes, wed me. He went away – I had nursed him when he was sick – but he never returned. Then I saw that the gods of the Kerlistians [Christians] lied, and I went back to my own people.[61]

But Lisbeth also expresses her sexual desire for Kim. After she arranges for her husbands – she is polyandrous – to carry away the ailing Lama, she beckons Kim into her hut. Stooping over a cashbox, she tells Kim in words charged with sexual desire "Shall I show thee how the Sahibs render thanks?" And when Kim claims he was a "wandering priest," she repeats herself, saying: "*if* thou wast a Sahib, shall I show thee what thou wouldst do?"[62] Kim refuses this sexual advance by putting his arm around her waist and kissing her on the cheek. The authorial voice follows this up by commenting that kissing "is practically unknown among Asiatics, which may have been the reason that she leaned back with wide-open eyes and face of panic."[63] Kim's refusal of the sexual encounter demonstrates another instance of the narrative's insecurity about Kim's identity. If he has a sexual encounter with a native woman it would not be – as in the case of Lispeth's first partner – an interaction that retains the racial divide. Rather, and on the contrary, it would collapse the racial divide owing to Kim's location as an insider and an outsider. In addition, this refusal also functions as a test for Kim, demonstrating that he can "hold off" the temptation of going native.

This duality of identity affects Kim's subjectivity and he is shown questioning his sense of self over the course of the novel. At a central moment in the narrative, Kim asks himself "Who is Kim?" and this is followed up by the narrator noting that he "considered his own identity, a thing he had never done before."[64] A little later, Mahbub Ali himself acknowledges Kim's inherent schizophrenia when he says that Kim was a sahib among sahibs, before trailing off regarding his South Asian identity. And Kim asks Mahbub Ali: "What am I? Mussalman [Muslim], Hindu, Jain or Buddhist?"[65] Kim goes on to say that "In all India there is no one so alone as I" before exclaiming "Who is Kim – Kim – Kim?"[66] And finally, convalescing after his encounter with the Russian and the Frenchman, the question reappears to him. His attempt to answer the question leads him to ask: "I am Kim. I am Kim. And what is Kim?"[67] Kipling fails to provide a convincing answer, or for that matter, any answer to this question, but it is illuminating to see that Kim is referred to as a "who" and as a "what," that is to say, as a subject as well as an object. This dichotomy of Kim's subjectivity is never fully settled in the novel.

As a spatial extension of this problematic, the socio-spatial identity of British India – whether it is a *functional* space or a *lived* space – is not resolved either. Kim therefore is condemned to remain both an insider as well as an outsider to the space of India. This question of course is taken up and resolved in a radically different way by Tagore through his Irish foundling protagonist, Gora, in the eponymously titled novel.

Envisioning heteroglossia: Tagore's *Gora*

Gora is set in late-nineteenth century Bengal (and British India), and documents its protagonist's move from a parochial to a more universalist conception of India. Gora, the eponymous protagonist, is presented as a "young, passionate but scholarly social reformer who has turned ultra-Hindu. He believes that the humiliation of being colonized can only be overcome by a tough protectiveness towards everything indigenous."[68] Thus, he takes ritual dips and maintains caste prohibitions and distinctions, especially about commensality, not in search of piety or purity but to stand with the common people. Such is his resolve that he refuses to eat or drink in the room of his mother, Anandamoyi, because she has a Christian maid. Gora, moreover, is revealed to be an Irish orphan who was adopted by Anandamoyi and raised as her younger son. The narrative informs the readers, but not Gora, about his origins, rendering Gora's attempts to be an orthodox Hindu deeply ironic. It is worth noting that the word *gora* [গোরা] that Tagore uses for his protagonist means "fair-skinned" in Bengali and is used to refer to Europeans and races considered "white." More crucially, the Bengali word [গোঁড়া] – a nasal variant of the proper noun Gora – also means orthodox or dogmatic, and Tagore's use of the word appears a deeply conscious choice.[69] The novel is organized around a series of events that constitute attacks on Gora's idea of India. These contending ideas of space and society are articulated in and through dialogue. As Supriya Chaudhuri notes:

> All the characters in *Gora* talk a great deal – it is mainly a novel of talk – and Gora himself, having graduated from anti-British, radical iconoclast to orthodox Hindu rebel in his effort to identify himself with his homeland, holds forth at length on his ideology and his goals.[70]

As Gora is challenged, he is forced to look for more authentic forms of patriotism and faith, and thus, the novel narrativizes the education of his counter-colonial spatial desire. Gora's transformation begins during his travels in the countryside where he encounters the anti-colonial resistance offered by a predominantly Muslim village. He also witnesses the Hindu village barber practising a more open Hinduism and living in solidarity with the Muslims rather than the Hindu landlords thus prompting him to rethink his understanding of the Indian community.

Gora is set in late-nineteenth-century Bengal when it was witnessing a number of indigenous reform movements aimed at purging Hinduism and Hindu society of its "social evils." It was also a time of the rise of Hindu reactionary orthodoxy in the face of colonial rule and extensive missionary proselytizing in British India. Bengal – and British India – at this time was very much a divided space with social fault-lines separating the orthodox Hindus from more heterodox ones. Key among these movements were the Brahmo Samaj, established by Tagore's father Devendranath, that became a major institution advocating the reform of Hinduism in nineteenth-century India.[71] Since 1830, especially in opposition to British attempts to ban the practice of sati or widow-burning, the orthodox Hindus of Bengal, led by Bhabanicharan Bandyopadhyay, organized themselves under the banner of the Dharma Sabha.[72] They sought to retain what they thought was essential about Hindu identity, namely, ritual practices and caste purity through the maintenance of caste distinctions and untouchability. The Brahmo Samaj, on their part, differed from mainstream Hindu faith in a key manner. They were opposed to the worship of idols. The members of these sects held varying views on colonial rule but they were fundamentally antagonistic to each other. They avoided social contact, often refusing to partake of food in the other's company, and in a bid to maintain the purity of their positions, forbade marrying across the sectarian divide. This divide between Hindu orthodoxy and the Brahmo reform movement forms the backdrop of *Gora*. And while Ashish Nandy opines that this ideological debate may make "no sense to the non-Bengali reader or to Bengalis of later generations," it is precisely the clash of the orthodox and heterodox worldviews that makes the novel tellingly relevant, and remarkably contemporary, to our present moment in India and beyond.[73] In the course of the novel, Gora meets Poresh Bhattacharya (Poresh Babu) because of his friend Binoy. Poresh Babu is portrayed as an open-minded Brahmo but married to the Anglicized Baradasundari, who is "as particular about maintaining the purity of Brahmoism as Gora is of Hinduism."[74] Towards the end of the novel, Gora realizes that he is racially Irish and thus out of the pale of orthodox Hinduism. But unlike Kim, whom Kipling projects as "white" and an "outsider" to India, Tagore pushes this towards a utopian conclusion suggesting that being unmarked by distinctions of caste or community makes Gora more of an Indian not less so. The novel has a radically utopian denouement: its end suggests the possibility of a marriage between Gora and Sucharita, Poresh Babu's foster daughter, and thus gestures towards possibly the *only* bi-racial marriage plot in a Bengali novel from the colonial era.

The novel opens in Calcutta and the narrative in the initial stages seems rooted in various indoor locations, with its most immediate reference for community being the colonial Bengali elite classes. The novel's opening stages a chance encounter in the streets of Calcutta that brings several key figures of the novel into contact with each other. In the opening pages of the novel, Gora's friend Binoy gazes out on to the Calcutta streets to see a carriage and a hackney-cab collide with each other. As Binoy goes out to help, he brings the injured passenger Poresh Babu and his foster daughter, Sucharita, into his home and calls for a doctor to attend to him. Tagore appears to play with the idea of the "accident" as both *mishap* and *chance* to get these characters, all central to the narrative's progression, to come into contact with each other. This also begins the process of Binoy, and eventually of Gora, interacting with Poresh Babu's family and coming in contact with the Brahmo ideologue Panu Babu. By putting these characters in contact with each other, the novel shows the Bengali elite community from the colonial era divided along Brahmo and Hindu lines. This gives readers a sense of the lived life in the extended Bengali family from the late-nineteenth century drawing out successfully, and realistically, the elite formation of colonial Bengal with its attendant divisions and biases. But the encounter also points to another crucial aspect of the novel: of the outside (the accident in the street) quite literally moving into the home (in the figure of Poresh Babu and Sucharita) and eventually structuring it (the marriage of Binoy and Lolita; and Sucharita and Gora). The utopian charge of the novel – and this is true of much of Tagore's novelistic oeuvre as well – lies in showing the *indoor* or the *local* as not a static or closed entity that is hermetically sealed and cordoned off from *the world* beyond. Rather, it shows the scale and the site of the local as one that is constantly made and remade by larger historical processes.[75]

The novel, in its initial stages, deploys the local as a means of conveying an idea of the socio-spatial community. The characters are often placed within the confines of their houses and shown discoursing about India, questions of caste and gender as well as other matters concerning Hindu orthodoxy and heterodoxy. However, the focus on the local is not without an awareness of the larger dimensions that structure it. Take, for instance, the opening of the second chapter:

> The dark skies seem to have been made wet and heavy by the monsoon evening. Under the silent rule of the colourless and unvarying clouds, the city of Calcutta [কলিকাতা] lay *like a giant, dreary dog that had curled up quietly with its head in its tail* [একটা প্রকান্ড নিরানন্দ কুকুরের মত লেজের মধ্যে মুখ গুঁজিয়া কুণ্ডলী পাকাইয়া চুপ করিয়া পড়িয়া আছে]. It had been drizzling since the previous evening. And while it had made the road muddy, the drizzle had not shown the strength to wash away that mud. The rains stopped at four in the afternoon but the clouds continued to spell trouble. At that time – when the lonesome room offers no comfort at evening because of the impending rain and there is no relief out of doors either – two young men could be seen sitting on . . . the damp terrace of a three-storeyed house.[76]

The narrative details this scene with great care and begins with the skies that have been "made wet and heavy" by the monsoon evening to present the city of Calcutta through almost a long-tailed simile: like a "giant, dreary dog that had curled up quietly with its head in its tail." But the description of the sky and the city, of the rain falling on the road and turning it into mud moves eventually to the "damp terrace," and then to the figures of Gora and Binoy. The passage generates an overarching perspective that weaves together a worm's eye view of the sky, and a bird's eye view of the ground before turning the focalization on to the two figures on the liminal space of the terrace that is neither the ground nor the sky. It points to the kinship between Binoy and Gora and shows them out on the terrace because "the lonesome room offer[ed] no comfort at evening." The passage positions these characters in relation to, and shows their interaction shaped by, multiple spatial and temporal scales: the space of the house and the city, the sky and in the context of the time of day, and the season of the year. It shows the local (the terrace, the characters of Binoy and Gora) in relation to, and structured by, the scales that go beyond thus hinting at the interpenetration of these scales. This aspect of the novel – the penetration of the local with the global – finds its most crucial expression in the novel in the figure of Gora. He is an Irish foundling who is adopted into a Bengali *bhadralok* upper-caste family and comes into the intimate space of the Bengali family owing to the larger historical processes of British imperialism and the 1857 Uprising.[77] This allows the novel to posit a utopian horizon of the socio-spatial entity of India as a dynamic and interactive space.

The novel's focus on the local allows it to present it as a microcosm of the larger world. Thus, it shows not just the sectarian fissures of society at the level of the family but also takes up the "woman question" in colonial India. In the middle of the narrative, Binoy critiques Gora's vision of India for not being holistic enough. He says:

> our love for our country is significantly lacking. We only see half of India. . . . We see India as a land of men alone. . . . I can say this about you [Gora] that you do not think of women for a moment – you know the country as one without women. This cannot be the correct way of knowing the country.[78]

Gora responds by saying that Binoy's desire to see women everywhere, like Englishmen do, suggests an imbalance in his outlook. But he cannot dismiss this easily and Binoy presses on:

> You said what you did only to deceive yourself. We do not really see them, when we see the all too familiar image of women engaged in household work around the house. If we could only see them beyond the household needs that they fulfil, we would come to see the beauty and wholeness of our country – perhaps even see an image of the country for which it would be easy to give up our lives. At least we could never make the mistake of thinking that there are no women in this country. . . . And you will have to admit that because women are out of sight, our country has remained only a half-truth to us.[79]

Tagore uses Binoy's observations to provide a telling critique of women's unpaid work and the gendered division of labour in the colonial Indian social formation. And in so doing, he holds up the "woman question" in colonial India in its concrete form besides underlining the inherent sexism in the rhetoric of Indian anti-colonial nationalism, especially Hindu nationalism, that repeatedly stressed the need for developing a masculinist and militant ethos to protect the mother.[80]

But Tagore is also underlining the colonial social formation as a socio-spatial entity that was fissured along the lines of gender. In this regard, the female characters in *Gora* are presented as purposeful, agential women. This includes not only figures such as Lolita and Sucharita, as well as Gora's mother Anandamoyi, but also characters such as Baradasundari and Harimohini (Poresh Babu's widowed sister) towards whom the narrative has an implicit antipathy. In the field of Bengali literature, this marks a move away from the idealized women found in the romances of Bankim Chatterji. Chatterji's female characters either embody ideas of the pristine and the natural (such as Kopalkundala) or are allowed agency to a limited extent, which is then circumscribed to make them secondary to their husbands (such as Kalyani in *Anandamath* or Prafulla in *Devi Chaudhurani*).[81] Tagore's heroines are likeable and sometimes not; they are sometimes kind, at times idiosyncratic, but always purposeful humans who speak their minds and act to get things done. In presenting the women as individuals and not types, these depictions indicate the emergence of realism in the Bengali literary sphere.[82] And crucially, by drawing out female characters that are not normative, Tagore appears to be working out possibilities of female personae in the aesthetic domain that would have been disciplined in reality. The point to note here is that Tagore appears less invested in showing the community *as it is* but rather to bend the realist mode to make it work for a utopian purpose, to point to the community *as it might become*.[83] This tension between the socio-spatial entity as it is and the entity as it might be – that is to say, the tension between the *present* form of the socio-spatial and its *subjunctive* form – lies at the heart of the imagined geographies structured by spatial desire. The most powerful and concrete articulation of the dialectical tension in the South Asian context between the socio-spatial *as it is* and the socio-spatial *as it may become* was, as we have seen in chapter 1, Chatterji's *Anandamath*. Tagore takes up the same problematic as Chatterji but his engagement and vision are not parochial and sectarian but instead tinged with a radical utopianism that, as we shall see, finds a fuller expression in the writings of Hossain.

Tagore's novel also brings into view the underclass that is excluded from narratives by novelists such as Chatterji or turned into the fabulous by Kipling in *Kim*. Kim is the underclass in that novel but he does not really encounter any hardships and, as such, leads what is almost a mythical life. Tagore's brings out the grotty realities of life in the colony, most notably through the figure of Nanda. Gora, the novel tells us, visited the low-caste people of his neighbourhood every morning, of whom Nanda was his greatest admirer. Nanda was a carpenter's son who worked in his father's shop but he was also an accomplished marksman when it came to shooting birds and a good cricket bowler. Going to Nanda's house after a gap of a few days, Gora finds that he has died from tetanus after injuring himself with his chisel. He is told that Nanda's father had "wanted to call a doctor but his mother had kept insisting that their son had been possessed by an evil spirit" and thus hired an exorcist to chant mantras, beat the boy, and sear his body with hot rods.[84] Gora uses Nanda's fate to point to the fact that no matter how educated or "liberated" the upper classes may consider themselves to be, they cannot separate themselves from the rest of the social formation. "the mast of the ship cannot be without a care in the world," he tells Binoy, "if there is a hole in the hull."[85]

The novel also delineates the image of the Muslim as an underclass, which is remarkably different from the stylized image of that community as evil that we have seen in Chatterji's *Anandamath*. Calcutta's traffic allows Tagore to narrate a scene in which an aged Muslim carrying a basket of provisions on his head tries to cross the street and is almost run over by a carriage. It also allows the author to show the antipathy of rich, upper-class Bengali Hindus towards Muslims – especially Muslim peasants – drawing out the image of the British Indian social formation as fissured, uneven, and fundamentally heteroglossic. Tagore writes:

> the *babu* called out to him to move away from the path of the carriage. But the old man not having heard it, was nearly run over. He managed to save himself somehow but the contents of his basket were scattered all over the road. The *babu* in the coachbox was angry – he called the old man "damn swine" ['ড্যাম শুয়ার'] before striking him in the face with a whip that raised a bloody welt on his face.[86]

Gora is angered and moved by the old man's plight and tries to chase the carriage but comes back to help the old man. To this the old man protests saying, "why are you taking the trouble *babu* – these things can no longer be used."[87] Note how Tagore deploys the term "babu" – a class marker that also signifies male and upper-caste status – in both the old Muslim's address to Gora as well as in his depiction of the man in the carriage. This situates Gora, the man in the carriage, and the old Muslim man in specific class locations within the colonial Bengali social formation. At the same time, Gora is presented as one unlike the other, more beastly, members of his class. And though his attempts to help the Muslim man, though well-meaning, are futile, they nevertheless come across as strikingly radical for the late-nineteenth century. This is not only because most upper-class and upper-caste Hindus would refuse to help a lower-class Muslim man, they would also not want to be touched by him. By showing a person who considers himself an orthodox Hindu break the taboo of touch, and that too for a member of the Muslim underclass, Tagore is portraying a situation here not as *it would have been* but rather as *it could have been*. And in so doing, he is once again encoding his realistic depiction with a utopian projection. But there is a further aspect of Tagore's realism that should also be appreciated: in making visible the Muslim underclass in his novel, he is also sketching a realistic Muslim character. This *humanizes* the figure of the Muslim and moves away from their stereotypical depictions as rapacious tyrants as we have seen in *Anandamath*.

The issue of a fissured community and the figure of the Muslim return when Gora visits the villages of Bengal. It was for the first time, the narrator notes, that "Gora was seeing what India was like beyond the elite and affluent circles of Calcutta. This solitary and vast rural India: so insular, bigoted, and weak – and so completely unaware of its own strength; so ignorant and apathetic to its own welfare."[88] Through his travels in rural Bengal, Gora discovers that the idea of serving the country was much more complex than he had imagined. If he had expected the villages to be antithetical to the colonial alienation of the city, he finds instead villagers' inability to join hands and work together. In the case of a fire that broke out in one of the villages, Gora notes that there was "much shouting, running and wailing but very little structured action."[89] He realizes that he had, in fact, "romanticized the oppressed masses, and he had expected to find in them a source of strength; he had expected to find in Bengal's villages the real India."[90] In narrating Gora's romantic conception of rural Bengal (and India) as well as his subsequent transformation in the novel, Tagore anticipates a set of totemic positions on the Indian village through the decade of the 1920s to the 1940s that would be taken by major Indian nationalist leaders Mohandas Gandhi, Jawaharlal Nehru, and Bhimrao Ambedkar. Gandhi, for instance, would draw on the writings of Henry Maine to idealize India's villages saying his vision of "Village Swaraj [self-rule] is that it is a complete republic, independent of its neighbours for its own vital wants, and yet interdependent for many others in which dependence is a necessity."[91] Gandhi's idealizations of the Indian village would be much less enthusiastically accepted by Nehru and Ambedkar. In his *Discovery of India*, Nehru would suggest that "autonomous village community" was an important aspect of the "old Indian social structure" along with caste and the profoundly patriarchal joint family system.[92] But it would be Ambedkar who would dismiss Gandhi's views most strongly, especially during the Constituent Assembly debates:

> The love of the intellectual Indians for the village community is of course infinite if not pathetic . . . I hold that these village republics have been the ruination of India. I am therefore surprised that those who condemn Provincialism and communalism should come forward as champions of the village. What is the village but a sink of localism, a den of ignorance, narrow-mindedness and communalism?[93]

In the novel, Tagore underlines the foolhardiness of the venture of "liberating" India while remaining dogmatic and sectarian at heart. Indeed, the novel dramatizes the sentimental, as well as the political, education of Gora. The most crucial moment in Gora's trajectory of self-awareness is in another village where he comes across a Hindu family that has adopted a Muslim boy as their son. Gora is hungry and thirsty from his travels but cannot eat at their house because the presence of the Muslim boy violates his idea of ritual purity. When Gora chides the Hindu barber for breaking social custom by adopting a boy from another religious community, the barber tells him "we say Hari [Krishna], they say Allah – there is no difference."[94] Since readers know of Gora's Irish parentage and his own history of being adopted into the Bengali family, a knowledge that he does not possess, it imbues this exchange with a sense of dramatic irony. The barber narrates to Gora how the colonial state and the Hindu landlord oppress the peasantry. The boy's father, Foru Sardar, was an indigo farmer who, along with other farmers, clashed repeatedly with the landlord and the European indigo-planters over cultivation rights. The farmers wanted to cultivate rice while the planters insisted they grow the indigo plant, a cash crop. That year the farmers had "managed to raise a paddy crop on the freshly silted levees of the river. But about a month earlier the manager of the indigo factory came with a band of armed goons and forcibly harvested the grain."[95] In the ensuing melee, the boy's father Foru Sardar had hit the manager so hard on his right hand that he had had to have his arm amputated. Following this incident

> police oppression spread across the neighbourhoods like wildfire. The tenants were unable to keep any possessions at home, and women were harassed to the point of being dishonoured. Foru Sardar and several others were in jail, and a large number of villagers had fled. Foru's household was without food; his wife could not appear in public as the only sari she owned was in tatters. Their only son, the boy Tamiz, had always called the barber's wife "aunt" – she has brought him to their house and was looking after him as he had nothing to eat at home.[96]

This depiction underlines the violence of colonial rule and its resultant dispossession that was felt keenly in the nineteenth-century Indian countryside. Tagore is also, just as crucially, showing the affinity between the Hindu landlords and the colonial state. At the same time, by demonstrating the solidarity between the Hindu barber and the Muslim farmer, Tagore indicates that people can, and do, break ossified social customs such as untouchability, even if it was a conceptual impossibility for the urban Bengali *bhadralok*, thus pointing to the utopian horizon of being in the colonial Indian social formation. Tagore's depiction of the real conditions of existence in the Bengal countryside underlines how the farmers and the lower classes, irrespective of their religious beliefs, were oppressed and exploited by the colonial state and the upper classes (a process that continues in our contemporary postcolonial moment). And here he is once again moving away from presenting an image of the Bengal countryside as stereotypically idyllic, showing it instead to be a site of life and struggle.

After hearing the story, egged on by his other orthodox Hindu companions, Gora leaves the barber to look for a Hindu house where they can get food and water. The nearest Hindu house turns out to be that of the factory manager, Madhav Chatterjee, who had been tormenting the farmers. At this point, Gora turns back and heads for the barber's house and Tagore outlines his protagonist's thoughts for his readers:

> Hunger and thirst were overpowering Gora but the more he thought of having to safeguard his caste by eating at the house of the criminal tyrant Madhav Chatterjee, the more intolerable the thought became . . . "By making purity an external matter, India has been terribly unrighteous," he thought. My caste identity will remain pure if I eat in the house of one who brings trouble and oppresses poor Muslims but will be violated in the house of the oppressed who risks social censure to give shelter to the son of a Muslim![97]

Gora returns to the house of the barber and though he cooks for himself, decides to stay at their house. This marks his first break with orthodoxy that will finally culminate in his knowledge of his Irish provenance. In narrativizing Gora's gradual development from being a person trapped in the received ethos of his "culture" to someone who is at home in the world, Tagore appears to be alluding to a trajectory of critically understanding the self which takes place "through a struggle of political 'hegemonies,' from opposing directions . . . in order to arrive at . . . a higher level of one's own conception of reality."[98]

In the end, the image of India that *Gora* draws out is one that is profoundly fissured and fractured but which is also, nevertheless, a contingent and contradictory whole. Crucially, it is a socio-spatial totality which speaks in many voices, that is, in other words, a fundamentally heteroglossic entity. The novel, as we have seen, begins in Calcutta and is largely located in conversations and living rooms, that is to say, indoors. But it also ventures beyond the elite confines of the city, to bring into its ambit the underclass, members of the lower castes, as well as Muslims. And it moves beyond the city as well, to the Bengal countryside, to provide a sense of life that is lived beyond urban locations. It shows British India as a site of dwelling that is not restricted to any one class, transcending barriers of class, community, caste, and religion. This is perhaps best articulated in the resolution of the novel when Gora realizes that he is not really a Hindu but an Irish foundling who was adopted into a Bengali family. Tagore seems to suggest that it is precisely by losing his Hindu identity that Gora becomes truly Indian. In what appears to be a testament to the utopian horizon of the novel, Gora says:

> That which I sought day and night to become but could not, I have become that today. Today I am Indian [ভারতবর্ষীয়]. In me, there is no discord between any communities – Hindu, Muslim or Christian. Today, all the castes of India [ভারতবর্ষ] are my caste; the food they eat is my food. I have visited many districts of Bengal and accepted the hospitality from the lowest caste of villagers . . . but I never could really be with them. There has been an invisible distance between us that I had not managed to cross. And this gave me a sense of emptiness. I have tried to disavow this emptiness; even tried to beautify it with decorations. All this because I love India [ভারতবর্ষ] more than my life – and I could not tolerate even the slightest criticism of the portion of it that was visible to me. But today I am relieved that I have been delivered from those futile attempts at decorating my emptiness.[99]

By making Gora Indian at its resolution, the novel makes the *world* a part of the *home*. As we have seen in the case of the patriotic lyrics earlier, the novel too articulates a worldview where the foreign is made both *familiar* as well as *familial*. By this move, the socio-spatial entity that is India is turned into an open space to which anyone can belong. This forces the breaking down of identitarian boundaries, gesturing towards the utopian vision of the world as one community where the space of India simply anchors one set of these global relationships. Tagore's vision pushes India beyond its limits to make it expand and merge with the world. Unlike Kipling, who demonstrates a colonizer's anxiety of a collapse in the colonizer-colonized distinction, and therefore that of the metropole and colony, Tagore not only champions, but actually valorizes such a breakdown. Moreover, if we think back to Strachey's comments from the introduction about there being no such country called India, Tagore appears to be taking that colonial logic and pushing it to its logical extreme. If it is merely a geographic expression, then anyone who resides in it is an Indian, across the axes of class, caste, religion, and ethnicity.

The novel signals its conception of world-as-community-as-family through the hint of the possibility of a marriage between Gora and Sucharita, the only instance of a bi-racial marriage in the Bengali novel from colonial India. The possibility of a marital union between Gora and Sucharita signals a breaking of spatial and racial boundaries that simultaneously inscribes the institution of marriage with a utopian ideal. Needless to say, this interaction between the Irish and the Indian (and Bengali) is fundamentally different from Kipling's narration of the interaction between Lispeth and Kim. Another key figure from the novel who anchors its expansive vision is Gora's mother, Anandamoyi. She decides to adopt Gora after the 1857 Uprising and through her action, sets up the condition to develop and sustain the argument of world-as-community-as-family. Anandamoyi's act of adopting Gora resonates with the decision of the Hindu barber's wife to adopt a Muslim son. In both instances, the women are depicted as having decided on the adoption, according them a degree of social agency that was nearly unimaginable at that time, in life or in fiction. The novel depicts women not just as agents of social change but also of a transformative utopianism in South Asia. Tagore depicts women, whose domain in the nineteenth century typically was the domestic sphere, as agents of progressive change. And just as crucially, he is illuminating the domestic sphere as a possible site for radical action and utopian transformation.

Utopian intimations: Hossain's *Padmarag*

Hossain's *Padmarag* narrates the story of the novella's eponymous heroine Padmarag (who is also known as Siddika and Zainab). The novel opens with a man in western attire at the Naihati railway station who appears to be in a state of distress. After walking aimlessly around the station for a while, he meets three women and requests them for help to lodge his sister for two weeks. They suggest he send his sister to Tarini Bhavan, a home for widows and destitute women in Calcutta run by Dina-Tarini Sen, the widow of the lawyer Tarinicharan Sen. The sister, Siddika Khatun, reaches Tarini Bhavan, and Mrs. Sen (as Dina-Tarini is called by the residents and her co-workers) calls her Padmarag, the "ruby with the lotus hue."[100] Siddika stays on at Tarini Bhavan as her brother does not come to get her after the initial two weeks as he had promised, and is trained in a range of practical skills such as sewing, preparing medicines, cooking, typing, and eventually, nursing. She comes across as a reclusive person to those around her and does not speak of her past; when asked, she says, "I am all alone in this world with not a soul to call my own."[101] This sense of singularity also gestures towards her similarity with Kim and Gora.

Over time, Siddika goes from being a loner to bonding with Tarini Bhavan's older residents. As she becomes acquainted with other residents, she realizes that they too have "suffered untold misery before finding their vocation at the institution."[102] In time she begins to feel a sense of belonging to the community of the workers and residents of Tarini Bhavan. And eventually in the course of the narrative, we also hear her story: her real name is Zainab and she was brought up in Chuadanga by her brother Muhammad Suleiman. She was betrothed to Latif Almas but the marriage could not take place as the prospective groom's uncle wanted Siddika's family to write away the property in her name to the groom before the marriage. Readers are also told of the land dispute between Suleiman and the British landlord and planter Charles James Robinson. This resulted in the British landlord murdering Suleiman and his elder son after failing in more insidious attempts at land-grabbing. Robinson had tried to frame Siddika for the murders who, after considering suicide, fled to Calcutta from Chuadanga. En route she had missed the connecting train at Naihati where she – disguised as a man and impersonating her brother – encountered, and sought help, from the three women who directed her to Tarini Bhavan. Here Siddika's disguising of herself as a man, to find a place for herself in a man's world recalls the cross-dressing heroine of *Anandamath*, Shanti. But unlike Shanti, who advances a militant Hindu ethos in Bengal against the Muslims, Hossain's vision of her heroine, as well as the socio-spatial entity of Bengal is radically inclusive. The novella underlines Siddika's coming into her own with the help of women from communities: Bengali Hindus, Brahmos, Bengali as well as Urdu-speaking Muslims, and Christians. This makes the socio-spatial vision of *Padmarag*, and its delineation of the postcolony, fundamentally different from Chatterji's.

In a somewhat contrived twist to the narrative of *Padmarag*, Siddika encounters both Latif and Robinson, though separately, while travelling with her co-workers. While travelling to Ranchi she finds Robinson grievously hurt from a riding accident. After regaining consciousness, Robinson expresses contrition for his actions towards Suleiman and Zainab (Siddika). Moved by his condition, Siddika reveals her identity to him, and Robinson seeks her forgiveness before passing away. Earlier in the narrative, Siddika and her companions find an injured Latif lying by the road in Kurseong. They nurse him back to health and eventually Latif realizes that Siddika is Zainab, the woman he was supposed to marry. The latter half of the narrative is largely focused on the question of their marriage. Latif asks Siddika to be his wife but she chooses an unusual path for herself. In a show of agency that would have been considered scandalously radical at time of *Padmarag's* publication, Siddika refuses Latif's proposal saying, "No. You find your way; I shall find mine."[103] Hossain portrays Siddika as a deeply agential woman who chooses to lead her life on her own terms. And her decision to refuse a married life powerfully underlines the feminist utopianism that is a hallmark of the text.

Hossain presents Tarini Bhavan as a spatialization of the feminist utopianism of *Padmarag*. This becomes evident in the manner in which the space is introduced:

> Dina-Tarini set up a home for widows. She named it Tarini Bhavan. Encouraged by its success, she established a school and formed a society called the Society for the Upliftment of Downtrodden Women. Located at one end of

> the huge mansion housing Tarini Bhavan was the girls' school; at the other end stood the home for widows. But as time went by, Dina-Tarini also felt impelled to found a Home for the Ailing and the Needy next to it.[104]

And while its founder was scorned by her various relatives for spending her wealth on this effort, the narrative makes clear the value of this initiative. We are told:

> Where would a widow find refuge, when she had no one in the world to turn to? In Tarini Bhavan [ভবন]. Where would a young orphan girl with no relatives be educated? At Tarini School [বিদ্যালয়]. Where would a wife go, when she was forced to leave her marital home because of her husband's intolerable brutality? To the same Tarini Workshop [কর্মালয়].[105]

It is pertinent to note the three distinct ways in which Dina-Tarini's initiative is spatialized in this passage. It is spoken of as "Tarini Bhavan" where "bhavan" [ভবন] literally means house, thus signalling the space as a home and shelter. It is also called "Tarini School" and the word used is "bidyaloy" [বিদ্যালয়], which literally translates to "abode of learning." And it is also presented as "Tarini Workshop," where the word used is "kormaloy" [কর্মালয়] that literally means "abode of work." By signalling Tarini Bhavan in this way, the narrative posits it as a space and one that brings together rest, learning, and work, becoming, in effect, a space of dwelling structured by labour and leisure; where thought, speech, and action are honed and actualized.

Tarini Bhavan is a space that trains women to be independent and self-reliant and, as such, is a site that actively seeks to undo the reproduction of patriarchal ideology. It is also, just as crucially, run by women. Tarini Bhavan is thus a space where women are in control of their own lives as well as the lives and livelihoods of others, including men. We are told:

> That evening, Dina-Tarini [Mrs. Sen] was very busy. A host of details pertaining to the school needed to be taken care of. She would have to serve as arbiter for the complaint that the collection-and-purchase clerk had, for some reason, beaten up the coachman. That afternoon, the female attendant who accompanied the schoolchildren travelling in Bus No. 5 had exclaimed, "Deary me! Never again will I escort a policeman's daughter home." The Nespector [Inspector] said, "I'll impound your bus, and toss your coachman into jail! . . . [because the] horse is lame."[106]

The image presented here is of a woman in the thick of taking care of administration, and in the running of an institution. This is particularly underlined when Siddika arrives in Tarini Bhavan and is taken to meet Mrs. Sen.

> a letter had arrived from the Society for the Prevention of Cruelty to Animals, stating that they would file a lawsuit against the school because the horses drawing Bus nos. 3 and 5 limped. A charge of theft had been levelled

> against two syces for siphoning off the feed purchased for the horses. That afternoon, a wheel from Bus No. 7 had broken off after the vehicle collided with the tram.[107]

These images may seem unremarkable and banal to contemporary readers. But it bears stressing that they are phenomenally egalitarian, and radical, for the first quarter of the twentieth century. Mrs. Sen is shown here not just taking care of the running of Tarini Bhavan but also dealing with what is implicitly suggested as the *unreasonableness* of some of those around her: the collection-and-purchase clerk beating up the coachman, and the female attendant refusing to escort a policeman's daughter in the future. In addition, she is also negotiating the demands of the law and the rules of the colonial state. It is therefore unsurprising that when Siddika comes to the Tarini Bhavan, Mrs. Sen simply does not have the time to meet her saying, "I won't be able to make inquiries about her today. We can meet early tomorrow morning and discuss the subject. Please leave me alone now – I am terribly busy."[108] Despite this passing mention, I find the depiction of Mrs. Sen as busy to be actually quite important. It shows her in control of her time and as one who sets her own priorities. Moreover, the idea of being busy and preoccupied is typically, and most certainly in the early-twentieth century, the sole preserve of men in a patriarchal set-up. By presenting her as "busy," and thus refusing to engage with the newcomer to Tarini Bhavan, the narrative claims for Mrs. Sen a privilege that is traditionally accorded to men. In so doing, it codes the everyday with the reversal of traditional gender norms, and therefore, it excludes a feminist utopianism. These extracts also point to the work that Mrs. Sen puts in for the running of her institution that is implicated not only in the lives of its residents but also keenly engaged with people (as well as animals) beyond. They gesture towards a dynamic link between the space of Tarini Bhavan and the social world beyond. Hossain appears to be suggesting that Tarini Bhavan is not a secluded space, or a space of seclusion, but one that is actively and vigorously engaged with the outside world. She appears invested in stressing, like Tagore in *Gora*, the dynamic inter-relationship between the *local* space (Tarini Bhavan) and the wider socio-spatial entity of which the local is constitutive, and a part. Furthermore, by situating Mrs. Sen at the helm of this interaction between the local and the social, *Padmarag* unequivocally champions the centrality of women not only in the life of society but also in directing social – and public – life. In effect, the novella re-imagines the place, and role, of women in South Asian society insisting that they are more than able to fulfil leadership roles.

Tarini Bhavan is also presented as the refuge, and training ground, of women rejected by patriarchal society. This is articulated ingeniously by Hossain through her use of the phrase "B.A. fail," which is directly referenced when Saudamini tells Siddika that "I failed my BA exam by only three marks [মাত্র তিনটি নম্বরের জন্য আমি বি. এ. ফেল কইয়াছি]."[109] The phrase "B.A. fail" points to, as Bagchi notes, Hossain's wordplay with B.A. (the Bachelor of Arts) as well as *biye* [বিয়ে], the Bengali word for marriage, conveying at once "the pathos of a failed marriage – the most common vocation of women in those days – as well as the failure to obtain . . . a college degree."[110] In addition, *Padmarag* provides spatial depictions that serve as analogues or counterpoints to Tarini Bhavan. It is a space of hope for the women who lived there, and the sense of hope is signalled through the stories that its residents tell. In telling the stories the residents such as Saudamini and Helen Horace signal a hostile world from which Tarini Bhavan gives them refuge. It is thus shown to be different from the world beyond. But other spatial depictions in the novella point to spaces of counter-patriarchal (and counter-colonial) possibility. Take, for instance, the passing reference to the waiting room of the Naihati railway station where we first encounter Siddika dressed as a man and impersonating her brother. The waiting room, as in *Anandamath*, brings to light Chakrabarty's idea of the "waiting room of history" and signals the idea of *passive waiting* instead of *active travel*.[111] This is contrasted with and implicitly made a spatial counterpoint to Tarini Bhavan, which becomes a spatialization of agency and action. And this contrast is paralleled in the figure of Siddika: in the first instance, she disguises herself as a man in order to be agential, as opposed to her in Tarini Bhavan where she becomes an agential subject by her own right, surrounded by other women like her. Hossain's repudiation of the waiting room motif through the delineation of a space like Tarini Bhavan marks her difference from the vision of Chatterji who, as I have shown in chapter 1, not only has gestures towards a future Hindu rule of the postcolony but also acquiesces to John Stuart Mill's logic of the colonized not being ready to rule themselves.

But perhaps the most moving image of the socio-spatial in the novel is not from within Tarini Bhavan but a depiction of the Bengal landscape. Hossain depicts the Bengal landscape as a lived space and home that is also – and here she is once again remarkably different from Chatterji – a space of Muslim dwelling. As Siddika walks about absentmindedly outside the Naihati station, we are told:

> Now, with the night waning, the moon's radiance paled. A light veil of darkness hung over the earth. A constellation of stars gently lit up the expanse of the sky . . .
>
> The melodious call to morning prayers from a mosque nearby awakened the silent earth, delighting in moulding it into the *rasa* of devotion. Hearing that call, nature aroused itself; all the birds awoke. The breeze carried the sound far, still further. That celestial sound was enough to make the traveller oblivious to the fatigue of having stayed up all night. I cannot describe what kind of joy he felt, since it can only be experienced; he who has experienced it, knows it.[112]

This passage is one of the depictions of space from beyond Tarini Bhavan, and it is also one of the more exuberant descriptions of space in the entire novella. It opens with Hossain presenting the space of Bengal spread out underneath the open sky and Siddika as a figure in that landscape. This landscape is powerfully marked with a sense of place, and this is done by inflecting the representation of space with a range of experiential registers. Notice how the passage, in particular, gestures to the interaction between the social world of humans and the natural world. The "call to morning prayers" from the mosque, which signals the social, engages the natural world, as it appears to *mould* the earth into "*rasa* of devotion." And this interactive tenor of social and the spatial is continued with the suggestion that the call to prayer animates nature and is carried further by the breeze. Moreover, the depiction suggests that the joy of the situation is indescribable: it can only be *experienced* turning it into a quasi-mystical event that is firmly located in the landscape. This is further underlined by suggesting that this entire moment has a salutary effect on the individual subject, namely, Siddika, for it restores her from her fatigued state. This image of the Bengal landscape as a lived space anticipates the other lived space that is so central to the narrative of *Padmarag*, Tarini Bhavan. Whereas the Bengal landscape is posited as a space of *contemplation*, Tarini Bhavan appears as world of *action* unto itself that is, just as crucially, run by women.

The depiction of Tarini Bhavan in *Padmarag* serves as a scaled-down version – a metonymic *stand in* – for the larger homeland of Bengal. But it is also a scaled-down version that is fundamentally transformed and made utopian: it is not a microcosm of Bengal *as it is* but of Bengal as *it might be*. Hossain, like Chatterji in *Anandamath* and Tagore in *Gora*, is also engaging with the utopian project of illuminating a possible *subjunctive* form of the socio-spatial that is utopian in relation to the present. And like Tagore, Hossain provides the image of a possible world that is dramatically different from the sectarian vision of Chatterji. Moreover, if Tagore depicts agential women in *Gora* and makes visible the Muslim underclass, Hossain takes that further in *Padmarag*. She makes visible not only Muslim Bengal in the abstract but also, through the figure and life-worlds of Siddika, the life of a Bengali Muslim woman in the concrete and her engagement with others in the social formation of colonial Bengal and India. While Hossain is rightly considered a pioneering feminist of colonial Bengal (and India), in *Padmarag* she highlights another aspect of her work that is just as significantly ground-breaking. The novella underlines Muslim claims to the space of Bengal (and India), and their role in shaping it, which are typically understated, if at all present, in canonical works of Bengali literature from the colonial period.

Hossain's *Padmarag* is also a significant literary registration of, and a literary engagement with, the "woman question" in colonial India. As such, it troubles the central claim made by Partha Chatterjee about the nationalist resolution of the woman question. In his discussion Chatterjee suggests that Indian male nationalists made the category "woman" synonymous with home, purity, and the inherent spirituality of the Indian nation.[113] These nationalists demanded that the British not interfere in this private world of home through social reform projects. The "woman question" for Chatterjee was resolved by Indian nationalists through a gendered mapping of the spiritual-material divide on to the space of the home [ঘর; *ghar*] and the world [বাহির; *bahir*] that became, respectively, the space for the nation's women and the men. And while the women's education is crucial for his argument, he suggests that this was structured by the creation of a new patriarchy by Indian nationalists and furthered the home-world distinction of Indian nationalism. Chatterjee's now canonical pronouncements on the women's question from colonial India and its subsequent resolution, however, omit other trajectories of the women's question. He does not engage at all, for instance, with the works of figures such as Hossain focusing instead solely on Hindu upper-class and upper-caste figures and their life-worlds. Chatterjee's archive is colonial Bengal, which he uses as a locus to make claims about nationalism in India and indeed, Asia and Africa. This is particularly curious as he does not seem to consider it worthwhile to examine any aspect of the majority community of colonial Bengal, that is, the Muslims, while making claims about that space. Furthermore, as Bagchi has noted:

> Chatterjee's account would see reformist women active in the field of women's education such as Rokeya Hossain or Pandita Ramabai, as already inscribed into the new patriarchy. However, the agency, critical questioning, and resourcefulness displayed by a number of colonial Indian women as actors in the public sphere shows that in fact there was no resolution of the woman's [sic] question in late-nineteenth-century or early-twentieth-century India, despite attempts on the part of the British colonial state and masculinist nationalism to do so.[114]

Hossain's *Padmarag* illuminates a striking trajectory of spatial desire that is directed at a profound egalitarianism across the axes of gender, class, religion, and race. Its utopian energies are directed at transcending patriarchal constraints and imagines a postcolony in, and through, the imagined geography of Tarini Bhavan. Hossain provides us not just with a feminist utopia but also a radical spatial image of egalitarianism that shows, in effect, a feminist resolution to the "women's question" in colonial India. While this undoubtedly was a minority view, and in many ways an impossible spatial desire, this *impossible* vision of egalitarianism and decolonization, nevertheless, gestures towards the extent of the *possible*. It lays out the utopian horizon of possibility and action to actualize a better and more just world.

Notes

1 Raymond Williams, *The Country and the City* (New York: Oxford University Press, 1973), 165.

2 Barnita Bagchi, *Introduction to Sultana's Dream and Padmarag: Two Feminist Utopias* (New Delhi: Penguin Books, 2005), xiii.

3 See Bharati Ray, *Early Feminists of Colonial India: Sarala Devi Chaudhurani and Rokeya Sakhawat Hossain* (New Delhi: Oxford University Press, 2002).

4 There are of course some notable exceptions to this. Sangeeta Ray discusses "Sultana's Dream" in Sangeeta Ray, *En-Gendering India: Woman and Nation in Colonial and Postcolonial Narratives* (Durham, NC: Duke University Press, 2000). Barnita Bagchi, in addition to editing "Sultana's Dream" and translating *Padmarag*, has also written on Hossain's literary oeuvre. Barnita Bagchi, "Ladylands and Sacrificial Holes: Utopias and Dystopias in Rokeya Sakhawat Hossain's Writings," in *The Politics of the (Im) Possible: Utopia and Dystopia Reconsidered*, ed. Barnita Bagchi (New Delhi: Sage Publications, 2012), 166–178. Gayatri Chakravorty Spivak refers to Hossain in passing

in her foreword to *Outside in the Teaching Machine*. See Gayatri Chakravorty Spivak, foreword to the Routledge Classics Edition, *Outside in the Teaching Machine* (New York: Routledge, 2009), xiv.

5 Mohandas Karamchand Gandhi, "The Great Sentinel," in *The Mahatma and The Poet: Letters and Debates Between Gandhi and Tagore 1915–1941*, ed. Sabyasachi Bhattacharya (New Delhi: National Book Trust, 1997), 93.

6 Bagchi, Introduction, ix.

7 The "Great Game" is a euphemistic term used to describe the strategic rivalry and conflict between the British and Russian Empires for supremacy in Central Asia during the late-nineteenth and early-twentieth centuries. See Peter Hopkirk, *The Great Game: On Secret Service in High Asia* (London: Murray, 1990), and Peter Hopkirk, *Trespassers on the Roof of the World: The Race for Lhasa* (Oxford: Oxford University Press, 1983).

8 Edward Said, *Culture and Imperialism* (New York: Vintage Books, 1994), 132.

9 Ibid.

10 Philip E. Wegner, " 'Life as He Would Have It': The Invention of India in Kipling's 'Kim,' " *Cultural Critique* (1993): 129.

11 Rudyard Kipling, *Kim*, ed. Maire ni Fhlathuin (Peterborough: Broadview Press, 2005), 53.

12 Ibid., 56.

13 Ibid., 54.

14 Ibid., 55.

15 Ibid., 58.

16 For an elaboration of the "simultaneity of the non-simultaneous," see Ernst Bloch, *Heritage of our Times*, trans. Neville Plaice and Stephen Plaice (Berkeley: University of California Press, 1990), 37–148.

17 For the idea of anachronistic space, see McClintock, *Imperial Leather*, 40–42.

18 Kipling, *Kim*, 73–74.

19 Ibid., 78.

20 Ibid., 185.

21 For an explanation of time-discipline under capitalist modernity, see E. P. Thompson, "Time, Work-discipline, and Industrial Capitalism," *Past and Present* 38 (1967): 56–97. Also see Georg Lukacs, "Reification and the Consciousness of the Proletariat," in *History and Class Consciousness: Studies in Marxist Dialectics*, trans. Rodney Livingstone (Cambridge: MIT Press, 1971), 83–222, especially "I. The Phenomenon of Reification."

22 For an elaboration of the denial of coevalness, see Johannes Fabian, *Time and the Other: How Anthropology Makes its Object* (New York: Columbia University Press, 1983).

23 Kipling, *Kim*, 79–80.

24 John Forbes Watson and John W. Kaye, "Preface," *The People of India: A Series of Photographic Illustrations, with Descriptive Letterpress, of the Races and Tribes of Hindustan: Originally Prepared Under the Authority of the Government of India, and Reproduced by Order of the Secretary of State for India in Council, Volume One*, eds. Watson and Kaye (London: India Museum, 1868). n. p.

25 Pinney, *Camera India*, 34.

26 Watson and Kaye (eds.), "Preface," *The People of India, Volume One,* n. p. My italics.

27 Kipling, *Kim*, 107.

28 Ibid.

29 Ibid., 111.

30 Ibid.

31 Watson and Kaye (eds.), "Preface," *The People of India, Volume Four,* plate 195.

32 Kipling, *Kim*, 111.

33 Ibid.

34 Williams, *The Country and the City*, 120.

35 Kipling, *Kim*, 112–113.

36 Ibid., 82, my italics.
37 Ibid., 95.
38 Ibid., 101.
39 Ibid.
40 Ibid.
41 Angus Wilson, *The Strange Ride of Rudyard Kipling: His Life and Works* (London: Secker and Warburg, 1977), 129.
42 Most readings of *Kim* elide the issue of the protagonist's Irish origin. Edward Said is a notable exception to this trend but he does not actually engage with the question in any great length. See Said, *Culture and Imperialism*, 132–161. Ian Baucom reads *Kim* as a "dramatization of the labors of imperial subject formation" but has nothing to say about his Irish identity. See Ian Baucom, *Out of Place: Englishness, Empire, and the Locations of Identity* (Princeton, NJ: Princeton University Press, 1999), 89.
43 Kipling, *Kim*, 94.
44 Ibid., 278.
45 Ibid., 284.
46 See Baucom, *Out of Place*, 86–100.
47 Kipling, *Kim*, 152.
48 Ibid., 170.
49 For colonial stereotypes such as "martial races," see Metcalf, *Ideologies of the Raj*, 66–159.
50 Kipling, *Kim*, 216.
51 Ibid., 277.
52 See Peter Heehs, *The Bomb in Bengal: The Rise of Revolutionary Terrorism in India, 1900–1910* (New Delhi: Oxford University Press, 1993).
53 Kipling, *Kim*, 274.
54 See Hopkirk, *Trespassers on the Roof of the World*, 20–54.
55 See Sandeep Banerjee and Subho Basu, "Secularizing the Sacred, Imagining the Nation-Space: The Himalaya in Bengali Travelogues, 1856–1901," *Modern Asian Studies* 49, no. 3 (2015): 609–649, 628–629. The Indians' achievements are downplayed in the otherwise masterful accounts of the Great Game by Peter Hopkirk.
56 Critics have conjectured if characters in *Kim* are based on real people from the British Indian Survey Department, particularly those who were involved in covert spying operations in Tibet. For instance, some have suggested Colonel Creighton is the literary avatar of Thomas Montgomerie, and Huree Chunder Mookerjee of Sarat Chandra Das. See Peter Hopkirk, *Quest for Kim: In Search of Kipling's Great Game* (Ann Arbor: University of Michigan Press, 1997).
57 Kipling, *Kim*, 163.
58 Ibid., 159.
59 Thomas Richards, *Imperial Archive: Knowledge and Fantasy of Empire* (London: Verso, 1993), 21.
60 See Rudyard Kipling, "Lispeth," in *Plain Tales from the Hills* (New York: Doubleday & McClure Co., 1899), 1–7.
61 Kipling, *Kim*, 297.
62 Ibid., 299.
63 Ibid. The idea of eastern sexuality as fundamentally "different" pervades not only writings from the colonial era but also more recent theoretical explorations of sexuality such as Michel Foucault's study on sexuality. In an orientalizing tenor, Foucault notes that the oriental "ars erotica" are fundamentally different from the "scientia sexualis" of the west. He notes, "China, Japan, India, Rome, the Arabo-Moslem societies . . . endowed themselves with an *ars erotica*. In the erotic art, truth is drawn from pleasure itself, understood as a practice and accumulated as experience; pleasure is not considered in relation to an absolute law of the permitted and the forbidden, nor by reference to a criterion of utility, but first and foremost in relation to itself; it is experienced as pleasure, evaluated in terms of its intensity, its specific quality, its duration,

its reverberations in the body and the soul. Moreover, this knowledge must be deflected back into the sexual practice itself, in order to shape it as though from within and amplify its effects." See Michel Foucault, *The History of Sexuality, Volume I: An Introduction*, trans. Robert Hurley (New York: Pantheon Books, 1978), 57. While Kipling and Foucault characterize "oriental sexuality" in differing ways – lacking pleasure in Kipling and overdetermined by pleasure in Foucault – both nonetheless engage in what can be understood as a copy-book form of orientalism.

64 Kipling, *Kim*, 163.
65 Ibid., 186.
66 Ibid., 226.
67 Ibid., 315.
68 Ashish Nandy, "The Illegitimacy of Nationalism: Rabindranath Tagore and the Politics of the Self," in *Bonfire of Creeds* (New Delhi: Oxford University Press, 2004), 184.
69 See Supriya Chaudhuri, "The Nation and Its Fictions: History and Allegory in Tagore's *Gora*," *South Asia: Journal of South Asian Studies* 35, no. 1 (2012): 106.
70 Ibid., 105.
71 See David Kopf, *The Brahmo Samaj and the Shaping of the Modern Indian Mind* (Princeton, NJ: Princeton University Press, 1979).
72 Bhabanicharan Bandyopadhyay was also the author of *Kalikata Kamalalay* (1823) that I have discussed briefly in my chapter on Calcutta.
73 Nandy, "The Illegitimacy of Nationalism," 184.
74 Ibid., 186.
75 This issue is taken up most directly in Tagore's novel *Ghare Baire* [ঘরে বাইরে] that is typically called *Home and the World* in its English translations. See Rabindranath Tagore, *Home and the World*, trans. Surendranath Tagore (London: Macmillan & Co., 1919).
76 Rabindranath Tagore, *Gora* [গোরা] (Calcutta: Vishwabharati, 2010), 13. All translations of passages from the novel cited here are mine.
77 For a comparative reading of *Gora* and *Kim* in the contexts of race and the 1857 Uprising, see Jaya Mehta, " 'Some Imaginary "Real" Thing': Racial Purity, the Mutiny, and the Nation in Tagore's *Gora* and Kipling's *Kim*," in *Rabindranath Tagore: Universality and Tradition* eds. Patrick Colm Hogan and Lalita Pandit (Madison: Farleigh Dickinson University Press, 2003), 199–212. For another comparative reading of *Gora* and *Kim*, see Lalita Pandit, "Caste, Race, and Nation: History and Dialectic in Rabindranath Tagore's *Gora*," in *Literary India: Comparative Studies in Aesthetics, Colonialism, and Culture*, ed. Patrick Colm Hogan and Lalita Pandit (Albany: State University of New York Press, 1995), 207–233. For another particularly creative reading of *Kim* and *Gora*, one that suggests that the latter has a lesson for upscale diasporics in globality in the contemporary moment, see Gayatri Chakravorty Spivak, "Resident Alien," in *An Aesthetic Education in the Era of Globalization* (Cambridge: Harvard University Press, 2012), 301–315.
78 Tagore, *Gora* [গোরা], 115.
79 Ibid.
80 See Sikata Banerjee, *Make Me a Man!: Masculinity, Hinduism, and Nationalism in India*. Albany: State University of New York Press, 2005.
81 See Chatterjee, *Kopal-Kundala*; and Bankim Chandra Chatterji, *Debi Chaudhurani, Or, the Wife Who Came Home*, trans. Julius Lipner (New Delhi: Oxford University Press, 2010).
82 In his essay "Bastob" [বাস্তব], which would translate as "Reality," Tagore comments in passing about his novel, *Gora*. He writes, somewhat sarcastically, that a critic "has charged that of all my writings, the one where all ingredients of reality has been assembled is the novel *Gora*." I find Tagore's comment to be less a repudiation of *realism* as an aesthetic mode than a rejection of the collapsing of the distinction between referent ("reality") and representation, and a concomitant insistence on the category of *mediation*. See Rabindranath Tagore, "Bastob" [বাস্তব], in *Sahityer Pothe* [সাহিত্যের পথে], Rabindra Rachanabali [রবীন্দ্র রচনাবলী], Vol. 23 (Calcutta: Vishwabharati, 1958), 364.

83 For an extended discussion on the utopian dimension of colonial realism, see Ulka Anjaria, *Realism in the Twentieth-Century Indian Novel: Colonial Difference and Literary Form* (New York: Cambridge University Press, 2012). It bears mentioning that Anjaria does not take up Tagore in her work in any substantial way but offers some tantalizing comments on character and temporality in Tagore's *Home and the World.*
84 Tagore, *Gora* [গোরা], 108.
85 Ibid., 109.
86 Ibid., 111.
87 Ibid.
88 Ibid., 183–184.
89 Tagore, *Gora* [গোরা], 184.
90 Pandit, "Caste, Race, and Nation," 229.
91 Mohandas Karamchand Gandhi, "Village Swaraj," *Harijan* 9, no. 28 (1942): 238.
92 Jawaharlal Nehru, *The Discovery of India* (New Delhi: Oxford University Press), 247.
93 Bhimrao Ramji Ambedkar, "Volume 7. 4th November, 1948" [7.48.231–7.48.232]. *Constituent Assembly of India Debates (Proceedings) – Volume VII.* Accessed November 15, 2018 from http://cadindia.clpr.org.in/constitution_assembly_debates/volume/7/1948-11-04
94 Ibid., 185.
95 Ibid.
96 Ibid., 186.
97 Ibid., 188.
98 Antonio Gramsci, "Notes for an Introduction and an Approach to the Study of Philosophy and the History of Culture," in *The Gramsci Reader: Selected Writings 1916–1935*, ed. David Forgacs (New York: New York University Press, 2000), 333–334.
99 Tagore, *Gora* [গোরা], 500–501.
100 Rokeya Sakhawat Hossain, "*Padmarag*," in *Sultana's Dream and Padmarag: Two Feminist Utopias*, ed. Barnita Bagchi (New Delhi: Penguin Books, 2005), 34. For the Bengali text of the novella, see Rokeya Sakhawat Hossain, *Padmarag* [পদ্মরাগ], in *Begum Rokeya Rachanabali* [বেগম রোকেয়া রচনাবলী], ed. Mostafa Mir (Dhaka: Bornayon, 2000), 261–368.
101 Hossain, *Padmarag*, 38.
102 Bagchi, introduction, xxi.
103 Hossain, *Padmarag*, 183.
104 Ibid., 27.
105 Ibid., 28.
106 Ibid.
107 Ibid., 28–29.
108 Ibid., 29.
109 Ibid., 104.
110 Bagchi, Introduction, xiv–xv.
111 Dipesh Chakrabarty, *Provincializing Europe: Postcolonial Thought and Historical Difference* (Princeton, NJ: Princeton University Press, 2000), 8.
112 Hossain, *Padmarag*, 24.
113 See the chapter "The Nation and its Women," in *The Nation and its Fragments: Colonial and Postcolonial Histories*, ed. Partha Chatterjee (Princeton, NJ: Princeton University Press, 1993), 116–134.
114 Bagchi, "Ladylands and Sacrificial Holes," 168.

Epilogue

That im/possible spatial desire called decolonization

In a discussion on utopia with Ernst Bloch and Horst Kruger, Theodor Adorno notes that

> insofar as . . . dreams have been realized, they all operate as though the best thing about them had been forgotten – one is not happy about them. As they have been realized, the dreams themselves have assumed a peculiar character of sobriety, of the spirit of positivism, and beyond that, of boredom.[1]

Adorno's comments highlight two important aspects of the utopianism that was decolonization. He notes that the attainment of the utopian ideal results in "a peculiar character of sobriety . . . and beyond that, of boredom," which surely points to the normative response to the term "decolonization," within the academe and beyond. Decolonization, the commonsensical understanding goes, is over; a matter of the past and of history. And just as crucially, Adorno also points out that "as these dreams have been realized, they all operate as though the best thing about them had been forgotten" thereby pointing to a *lack* that drives the utopian ideal and propels the continuation of the utopian quest. Utopia is the impossible horizon of action, quite like the desire for the pot of gold at the end of the rainbow. And indeed, it is akin to striving for the ever-receding horizon itself that looks as far even as we strive to move closer to it. But the articulation of, and the desire for, the *impossible* nevertheless draws out, and makes visible, the extent of the *possible*. But the actualization of a utopian ideal does not – ought not to – stall the utopian impulse or utopianism as such. To indicate as such would be to posit the end of history and that would reduce utopianism to a perverse teleology as laid out by, for instance, Francis Fukuyama.[2] Adorno's insights encourage us to think about the decolonizing imperative, and the utopian impulse structuring it, even as they sum up the utopian dialectic of decolonization perfectly: that the principle of hope actualized is always, and forever, intersected with the principle of "Not Yet."

For several generations of the wretched of the earth, decolonization had been a utopian ideal to unseat those who claimed themselves to be, to use Victor Kiernan's apt phrase, the lords of human kind.[3] And unseat them they did, through the course of the twentieth century making it a truly radical century, especially from the vantage-point of the twenty-first. And since decolonization is, as I have been suggesting in this book, a kind of utopian spatial desire, then it follows that the moment of decolonization is *both* the culmination of a certain structure of utopian desire and also the inauguration of it. It is *both* the principle of hope actualized *and* the expression of the spirit of "Not-Yet." And indeed, the global history of decolonization encourages us to think of it as both an *event* and a *process*. The idea of decolonization as an event is, in many ways, self-explanatory. It is the moment of the dismantling of the European political, economic, and cultural regimes. But the *moment of dismantling* is also always the *inauguration of the process of dismantling* these regimes. The understanding of decolonization as a process, which is undoubtedly the more challenging aspect of it, suggests that the transfer of political power is but a moment that inaugurates not just the dismantling of European structures of rule but also the reconstruction and reanimation of the political, economic, and the cultural spheres of the colonized, leading to the breaking of the "mind-forg'd manacles" and to decolonizing the mind.

The idea of decolonization, as simultaneously a moment and a process, is expressed most poignantly in the speech by Jawaharlal Nehru, the first Prime Minister of independent India, on the eve of Indian independence. Nehru moves the resolution of Indian independence in the British Indian Parliament on the midnight of 15 August 1947 with these words:

> Long years ago we made a tryst with destiny, and now the time comes when we shall redeem our pledge, *not wholly or in full measure, but very substantially*. At the stroke of the midnight hour, when the world sleeps, India will awake to life and freedom. A moment comes, which comes but rarely in history, when we step out from the old to the new, when an age ends, and when the soul of a *nation*, long suppressed, finds utterance. . . . At the dawn of history India started on her unending quest, and trackless centuries are filled with her striving and the grandeur of her success and her failures. . . . We end today a period of ill fortune and India discovers herself again . . . we have to . . . work hard, to give reality to our dreams. Those dreams are for India, but they are also for the world, for all the nations and peoples are too closely knit together today . . . this One World that can no longer be split into isolated fragments. . . . We have to build the *noble mansion of free India* where all her children may dwell.[4]

These words from Nehru's iconic "tryst with destiny" speech show his perceptive and direct engagement with the problematic of decolonization and its utopian character. The speech stresses the singularity of the event that is decolonization through a series of temporal markers. As one reads the words (or hears the recorded words being spoken), the temporal dimension of the speech immediately comes to our attention. It begins with the words "long years ago," and its invocation to years is followed by other temporal emphases, among them "now the time comes"; "the stroke of the midnight hour"; "the dawn of history," and "trackless centuries." These temporal markers stress the singularity of the event that was decolonization. And understandably so, for it was indeed a moment of keeping a "tryst with destiny" that sees the actualization of a certain kind of spatial desire. It was a moment, "a moment that comes but rarely in history," when a colonized subject people stepped out "from the old to the new" to redeem the pledge: to free themselves from the bonds of colonialism and actualize a better and more just world for themselves, for other subject peoples, and for the world at large. But as is evident from Nehru's speech, he does not simply stress decolonization as a *moment* of the transfer of political power but also acknowledges it as a *process*, drawing attention to the processual aspects of decolonization in diverse ways. If decolonization is a moment to "redeem our pledge," this excitement of redemption is tempered with the phrase "not wholly or in full measure." The notion of decolonization as process is adduced further when he states "we have to . . . work hard, to give reality to our dreams. Those dreams are for India, but they are also for the world."

Nehru's speech does not only underline the temporal dimension of decolonization but also underscores it as an actualization of spatial desire. This is highlighted in his statement that the politicians and people of India must now "build the noble mansion of free India where all her children may dwell." This sentence, in gesturing towards India as a "noble mansion" that requires to be built illuminates the logic of place-making that is so central to the decolonizing process. Note also his use of the world "mansion" and "dwell" that suggests India not as an abstract spatial entity but as a space of habitation and dwelling. But that is not the only way the spatial registers in Nehru's speech. He also notes that "all the nations and peoples are too closely knit together today" in this "One World that can no longer be split into isolated fragments." Nehru's invocation of the "One World" that is closely knit and not split into isolated fragments references the space-making dimension of colonial globalization that structured the world into bounded territorialized states. But the reference to the "One World" and the "closely knit" world does ring hollow given that British India was partitioned and the ethnic violence and displacement did indeed split South Asia into isolated fragments. Nehru would reference this more directly in his message to the press the following morning:

> The appointed day has come . . . and India stands forth again after long slumber and struggle, awake, vital, free and independent. . . . We think also of our *brothers and sisters who have been cut off from us by political boundaries* and who unhappily cannot share at present in the freedom that has come. . . . And to India, our much-loved motherland, *the ancient, the eternal* and the ever-new, we pay our reverent homage and we bind ourselves afresh to her service.[5]

His articulation of India as "the ancient, the eternal" exists in a paradoxical relationship to his keen awareness of the geographical expanse of the country brought about through the partition of British India into the independent dominions of India and Pakistan. And this is particularly evident in his reference to the "brothers and sisters who have been cut off . . . by political boundaries." In his speech and message, Nehru references two of the crucial spatial categories, and enduring fictions, that were forged in the crucible of colonial capitalist modernity: the *boundary* and the *nation*, which reciprocally inform each other. And both these categories would have a role to play in diminishing and arresting the utopian impulse that underwrote decolonization.

In the formerly colonized parts of the world, the spatial desire or structure of feeling that was decolonization crystallized into nationalism, and the postcolony was imagined as the horizontal comradeship that was the nation. In British India, the idea of the postcolony came to be understood as the Indian nation and subsequently, the Pakistani nation.[6] And nationalism, "that magnificent song that made the people rise against their oppressors" did indeed stop short, falter, and die away on the day that independence was proclaimed in South Asia.[7] In other words, the utopian impulse that underwrote decolonization was arrested, and absorbed into, the figure of the nation-state. And as a corollary, the so-called postcolonial states that governed the formerly colonized spaces implicitly accepted decolonization as an *event* of the transfer of political power from white to darker hands and de-emphasizing the idea of decolonization as a *process*. Instead of working towards a radical reconstitution of formerly colonized societies, that is, the "whole social structure being changed from the bottom up," the so-called postcolonial states made peace with the depredations of capital against labour and incorporated the colonial ethic into their outlook, thus taking a neo-colonial turn.[8] Most crucially, the world system engendered by colonialism – the one and unequal world of metropolitan and peripheral spaces now turned into a concatenation of nominally equal nation-states structured by the boundary-concept – was left untouched by the newly independent states. Instead, the territorialized state-space was valorized and coded with an affective charge and turned into a quasi-mystical category and a secular divinity. And the so-called postcolonial state in this milieu emerged as the sole arbiter of utopianism now recoded as progress, which was myopically understood from a strictly statist perspective.

These indicate, ultimately, to the failure of the utopian impulse of decolonization. The first intimation of this was probably the shock of violence that was directed by the colonized not at the colonizer but at each other. The Irish civil war at the dawn of Irish independence and the ethnic violence and displacement at the cusp of the colonizer's departure from South Asia, and the partition of British India, provide salutary reminders of that history. And this structure of violence in so-called postcolonial societies has continued and has typically been directed against "othered" communities residing with these spaces: women, working classes, religious minorities, immigrants, people considered to be of lower caste birth, of minority sexual orientation and gender identity, the First Nations and other autochthonous groups, to name just a few. In addition, colonialism and capitalism, while continuing to maintain their mutually affirming relationship, have re-invented themselves to deal with the "crisis" of decolonization. As the United States of America claimed the mantle of imperial metropole from the United Kingdom and France in a kind of modern *translatio imperii*, new modalities of rule ranging from coups and Napalm to debt bondage and outsourcing were developed to keep the formerly colonized spaces nominally free (though at times not) but always very far from their decolonizing missions. The failure of decolonization as a utopian project is underwritten by these twin emphases: of a reinvigorated US-led colonial capitalism across the globe and the abandonment of the utopian impulse of decolonization after the event by the so-called postcolonial states.

In post-independence India utopianism has been denuded of content and rendered into a purely formal category – a formula, if you will – that has been deployed, in particular, during the time of elections to promise people all but the moon. One Indian Prime Minister promised to provide the masses with "roti, kapda, aur makaan" [रोटी कपड़ा और मकान] or "bread, clothing, and housing" in the 1960s making that phrase a rallying cry for the elections. A decade later, having failed to provide all Indians with bread, clothing, and housing, the same politician now sought to remove poverty. Her newer utopian vision was articulated through the pithy imperative "garibi hatao" [गरीबी हटाओ] or "remove poverty" that, unsurprisingly, was not actualized either. The utopian ideal in the contemporary, and especially the so-called postcolonial, world has been reduced to empty sloganeering. The formulaic deployment of utopianism has continued in the twenty-first century with another Indian Prime Minister promising all Indian citizens "acche din" [अच्छे दिन] or "good days." He had memorably proclaimed "acche din aaney wale hain" [अच्छे दिन आने वाले हैं] – "Good days are coming" – after winning the elections in 2014. And as Indians, especially the socially and economically marginalized among them, wait for the promised good days to arrive, the din of "acche din" continues to only grow louder even as it appears, with each passing day, a close relative of Samuel Beckett's Godot. But it must also be noted that while the politicians undoubtedly make cynical use of these utopian slogans and promises, they nevertheless strike a chord with the people: perhaps because the utopian impulse – that principle of hope – is a crucial aspect of our lives.

A truly radical utopianism, one that transcends the formulaic utterances and leads to genuine social transformation, could surely do more than simply make noise and instead gain real traction in the postcolonial Indian social formation. And about this, there is perhaps hope. In 2016, the Indian government, clearly rattled by students raising slogans demanding independence from hunger, casteism, brahminism, religious sectarianism, and capitalism at a university campus in New Delhi, jailed them on charges of sedition, that is, waging war against the Indian state.[9] And in March 2018, about 50,000 Indian farmers – a social group that has been among the worst affected by the neoliberal restructuring of the India economy since the decade of the 1990s – marched 180 kilometres to Mumbai to protest against agrarian distress and demanding that the Indian state (not to mention, mainstream middle-class media outlets) take heed of, and address, their problems.[10] The utopian impulse that was decolonization may be dead, but the utopian impulse that is decolonization lives on nonetheless.

Notes

1 Ernst Bloch, Theodore Adorno, and Horst Kruger, "Something's Missing: A Discussion Between Ernst Bloch and Theodor W. Adorno on the Contradictions of Utopian

Longing," in *The Utopian Function of Art and Literature: Selected Essays*, trans. Jack Zipes and Frank Mecklenburg (Cambridge: MIT Press, 1989), 1.

2 See Francis Fukuyama, *The End of History and the Last Man* (New York: Free Press, 1992).

3 Victor G. Kiernan, *The Lords of Human Kind: Black Man, Yellow Man, and White Man in an Age of Empire* (Boston: Little Brown, 1969).

4 Jawaharlal Nehru, "Tryst with Destiny," in *Jawaharlal Nehru's Speeches. Volume One. September 1946–May 1949* (New Delhi: Publications Division, Government of India), 25–26. My italics.

5 Jawaharlal Nehru, "The Appointed Day," in *Jawaharlal Nehru's Speeches. Volume One. September 1946–May 1949* (New Delhi: Publications Division, Government of India), 27–28. My italics.

6 As I have pointed out throughout this book, the nation was not the only way the post-colony was imagined. However, the nation did indeed coalesce into the *normative* idea of the postcolony.

7 Frantz Fanon, *The Wretched of the Earth*, trans. Constance Farrington (New York: Grove Press, 1963), 203.

8 Ibid., 35.

9 See Priyamvada Gopal, "This is a watershed moment for India. It must choose freedom over intolerance," *The Guardian*, February 17, 2016. Accessed January 14, 2019 from www.theguardian.com/commentisfree/2016/feb/17/india-kanhaiya-kumar-watershed-freedom-intolerance-bjp-hindu

10 Rajendra Jadhav, "Thousands of farmers march to India's Mumbai to demand help," *Reuters US*, March 22, 2018. Accessed January 14, 2019 from www.reuters.com/article/us-india-farmers-march/thousands-of-farmers-march-to-indias-mumbai-to-demand-help-idUSKCN1GO09V

Appendix

Jana Gana Mana by Rabindranath Tagore

English translation of Jana Gana Mana

Translated by Sandeep Banerjee

Guide of people's thoughts, India's destiny you decree.
Glory be to thee!
The Punjab, Sindh, Gujarat, Maratha
The Dravid peninsula, Odisha and Bangla;
The Himalayan peaks, and the Vindhya,
The surging waves of the Yamuna and Ganga –
They rise in your name, seek to be blessed,
Singing you hymns of praise.
For you the people's well-being – glory to thee!
India's destiny you decree,
Glory, glory, glory to thee!

Hearing your welcome, heeding your call,
Hindu, Sikh, Jain and Buddhist, they all
With Muslim, Christian, Parsee meet,
To weave love's garland by your seat.
For you the people's unity – glory to thee!
India's destiny you decree,
Glory, glory, glory to thee!

Through fall and rise, on Time's rugged way
Your chariot-wheels ring out night and day
Amidst that mighty clamor, O eternal rider,
The sound of your conch-shell we hear.
From pain and distress, our savior!
You direct the people's paths – glory to thee!
India's destiny you decree,
Glory, glory, glory to thee!

Through the nights, dark and still
When this swooning land was ill
We had your lap, through visions and our terror
You watched and blessed us, our loving mother!
You quell the people's sorrow – glory to thee!
India's destiny you decree,
Glory, glory, glory to thee!

Now night gives way to the break of day,
The sun rays fall on the eastern hill
The singing birds, the blissful air,
Life with your joy and mercy they fill
India awakes, bowing to your feet,
Glory to thee, O King of kings
India's destiny you decree,
Glory, glory, glory to thee!

Bengali text of Jana Gana Mana

জনগণমন-অধিনায়ক
রবীন্দ্রনাথ ঠাকুর

জনগণমন-অধিনায়ক জয় হে ভারতভাগ্যবিধাতা !
পঞ্জাব সিন্ধু গুজরাট মরাঠা দ্রাবিড় উৎকল বঙ্গ
বিন্ধ্য হিমাচল যমুনা গঙ্গা উচ্ছলজলধিতরঙ্গ
তব শুভ নামে জাগে, তব শুভ আশিষ মাগে,
গাহে তব জয়গাথা।
জনগণমঙ্গলদায়ক জয় হে ভারতভাগ্যবিধাতা !
জয় হে, জয় হে, জয় হে, জয় জয় জয় জয় হে ॥

অহরহ তব আহ্বান প্রচারিত, শুনি তব উদার বাণী
হিন্দু বৌদ্ধ শিখ জৈন পারসিক মুসলমান খৃস্টানী
পূরব পশ্চিম আসে তব সিংহাসন-পাশে
প্রেমহার হয় গাঁথা।
জনগণ-ঐক্য-বিধায়ক জয় হে ভারতভাগ্যবিধাতা !
জয় হে, জয় হে, জয় হে, জয় জয় জয় জয় হে ॥

পতন-অভ্যুদয়-বন্ধুর পন্থা, যুগ যুগ ধাবিত যাত্রী।
হে চিরসারথি, তব রথচক্রে মুখরিত পথ দিনরাত্রি।
দারুণ বিপ্লব-মাঝে তব শঙ্খধ্বনি বাজে
সঙ্কটদুঃখত্রাতা।
জনগণপথপরিচায়ক জয় হে ভারতভাগ্যবিধাতা !
জয় হে, জয় হে, জয় হে, জয় জয় জয় জয় হে ॥

ঘোরতিমিরঘন নিবিড় নিশীথে পীড়িত মূর্ছিত দেশে
জাগ্রত ছিল তব অবিচল মঙ্গল নতনয়নে অনিমেষে।
দুঃস্বপ্নে আতঙ্কে রক্ষা করিলে অঙ্কে
স্নেহময়ী তুমি মাতা।
জনগণদুঃখত্রায়ক জয় হে ভারতভাগ্যবিধাতা !
জয় হে, জয় হে, জয় হে, জয় জয় জয় জয় হে ॥

রাত্রি প্রভাতিল, উদিল রবিচ্ছবি পূর্ব-উদয়গিরিভালে –
গাহে বিহঙ্গম, পূণ্য সমীরণ নবজীবনরস ঢালে।
তব করুণারুণরাগে নিদ্রিত ভারত জাগে
তব চরণে নত মাথা।
জয় জয় জয় হে জয় রাজেশ্বর ভারতভাগ্যবিধাতা !
জয় হে, জয় হে, জয় হে, জয় জয় জয় জয় হে ॥

Bibliography

Adorno, Theodore. *Negative Dialectics*. Translated by E. B. Ashton. London: Routledge & Kegan Paul Ltd., 1973.

Agathocleous, Tanya. "Imperial, Anglophone, Geopolitical, Worldly: Evaluating the 'Global,' in Victorian Studies." *Victorian Literature and Culture* 43, no. 3 (2015): 651–658.

Ahmad, Aijaz. *In Theory: Classes, Nations, Literatures*. London: Verso, 1992.

Ambedkar, Bhimrao Ramji. "Volume 7. 4th November, 1948" [7.48.231–7.48.232]. Constituent Assembly of India Debates (Proceedings) – Volume VII. http://cadindia.clpr.org.in/constitution_assembly_debates/volume/7/1948-11-04 (accessed November 15, 2018).

Amin, Samir. *Unequal Development: An Essay on the Social Formations of Peripheral Capitalism*. Translated by Brian Pearce. New York: Monthly Review Press, 1976.

Anderson, Benedict. *Imagined Communities: Reflections on the Origin and Spread of Nationalism*. London: Verso, 1991.

Anjaria, Ulka. *Realism in the Twentieth-Century Indian Novel: Colonial Difference and Literary Form*. New York: Cambridge University Press, 2012.

Ashraf, Kazi M., and Puran C. Joshi. *Rebellion, 1857: A Symposium*. New Delhi: People's Publishing House, 1957.

Assael, Brenda. "Beyond Empire: Globalizing the Victorians." *Victorian Literature and Culture* 43, no. 3 (2015): 643–650.

Atkinson, James. *The City of Palaces: A Fragment, and Other Poems*. Calcutta: Government Gazette Press, 1824.

Bagchi, Barnita. Introduction to *Sultana's Dream and Padmarag: Two Feminist Utopias*, by Rokeya Sakhawat Hossain, vii–xxvi. New Delhi: Penguin Books, 2005.

———. "Ladylands and Sacrificial Holes: Utopias and Dystopias in Rokeya Sakhawat Hossain's Writings." In *The Politics of the (Im)Possible: Utopia and Dystopia Reconsidered*, edited by Barnita Bagchi, 166–178. New Delhi: Sage Publications, 2012.

Bagchi, Jasodhara. "Representing Nationalism: Ideology of Motherhood in Colonial Bengal." *Economic and Political Weekly* 25, no. 42/43 (1990): WS65–WS71.

Baishya, Amit R. *Contemporary Literature from Northeast India: Deathworlds, Terror and Survival*. London: Routledge, 2018.

Bakhtin, Mikhail M. "Discourse in the Novel." In *The Dialogic Imagination: Four Essays by M. M. Bakhtin*, edited by Michael Holquist, translated by Caryl Emerson and Michael Holquist, 259–422. Austin: University of Austin Press, 1981.

———. "Forms of Time and the Chronotope in the Novel." In *The Dialogic Imagination: Four Essays by M. M. Bakhtin*, edited by Michael Holquist, translated by Caryl Emerson and Michael Holquist, 84–258. Austin: University of Austin Press, 1981.

Ball, Charles. *The History of the Indian Mutiny: Giving a Detailed Account of the Sepoy Insurrection in India: And a Concise History of the Great Military Events Which Have Tended to Consolidate British Empire in Hindostan*. London: London Print and Pub. Co., 1858.

Bandyopadhyay, Bhabanicharan. *Kalikata Kamalalay*. Translated by A. K. Roy. Calcutta: Firma KLM, 1990.

Banerjea, Surendranath. *A Nation in Making; Being the Reminiscences of Fifty Years of Public Life*. Bombay: Oxford University Press, 1925.

Banerjee, Sandeep, and Subho Basu. "Secularizing the Sacred, Imagining the Nation-Space: The Himalaya in Bengali Travelogues, 1856–1901." *Modern Asian Studies* 49, no. 3 (2015): 609–649.

———. "The City as Nation: Delhi as the Indian Nation in Bengali Bhadralok Travelogues 1866–1910." In *Cities in South Asia*, edited by Crispin Bates and Minoru Mio, 125–142. Abingdon: Routledge, 2015.

Banerjee, Sikata. *Make Me a Man! Masculinity, Hinduism, and Nationalism in India*. Albany: State University of New York Press, 2005.

Banerjee, Sumanta. *The Parlour and the Streets: Elite and Popular Culture in Nineteenth Century Calcutta*. Calcutta: Seagull Books, 1989.

Barrell, John. *The Dark Side of the Landscape: The Rural Poor in English Painting, 1730–1840*. Cambridge: Cambridge University Press, 1983.

Barrow, Ian. *Making History, Drawing Territory: British Mapping in India c.1756–1905*. New Delhi: Oxford University Press, 2003.

Basu, Subho. *Does Class Matter? Colonial Capital and Workers' Resistance in Bengal, 1890–1937*. New Delhi: Oxford University Press, 2004.

———. "The Dialectics of Resistance: Colonial Geography, Bengali Literati and the Racial Mapping of Indian Identity." *Modern Asian Studies* 44, no. 1 (2010): 53–79.

Bates, Crispin, and Marina Carter. *Mutiny at the Margins: Documents of the Indian Uprising*. New Delhi: Sage Publications, 2017.

Baucom, Ian. *Out of Place: Englishness, Empire, and the Locations of Identity*. Princeton, NJ: Princeton University Press, 1999.

Bautze, Joachim K. "The Beginning of Photography in India and the Photographers in *Picturesque Views*." In *Picturesque Views: Mughal India in Nineteenth-Century Photography*, edited by Raffael Dedo Gadebusch, 21–26. Ostfildern: Hatje Cantz, 2007.

Benjamin, Walter. "On the Mimetic Faculty." In *One-Way Street and Other Writings*, translated by Edmund Jephcott and Kingsley Shorter, 160–163. London: NLB, 1979.

———. *The Origin of German Tragic Drama*. Translated by John Osborne. London: Verso, 1998.

———. "The Work of Art in the Age of Mechanical Reproduction." In *Illuminations*, edited by Hannah Arendt, translated by Harry Zohn, 217–251. New York: Schocken Books, 1969.

Bernardini, Michele. "The *masnavi-shahrashub* as Town Panegyrics: An International Genre in Islamic Mashriq." In *Narrated Space in the Literature of the Islamic World*, edited by Roxanne Haag-Higuchi and Christian Szyska, 81–94. Wiesbaden: Harrassowitz, 2001.

Beverly, Henry. *Report on the Census of the Town of Calcutta Taken on the 6th April 1876*. Calcutta: Bengal Secretariat Press, 1876.

Bhabha, Homi K. "DissemiNation: Time, Narrative, and the Margins of the Modern Nation." In *Nation and Narration*, edited by Homi K. Bhabha, 291–332. London: Routledge, 1990.

Bhatia, B. M. *Famines in India; A Study in Some Aspects of the Economic History of India, 1860–1965*. Bombay: Asia Publishing House, 1967.

Bhattacharya, Sabyasachi. *Vande Mataram: The Biography of a Song*. New Delhi: Penguin Books, 2003.

Bhattacharya, Tithi. *The Sentinels of Culture: Class, Education, and the Colonial Intellectual in Bengal, 1848–85*. New Delhi: Oxford University Press, 2005.

———. "Tracking the Goddess: Religion, Community, and Identity in the Durga Puja Ceremonies of Nineteenth-Century Calcutta." *The Journal of Asian Studies* 66 (2007): 919–962.

Blake, William. "The Marriage of Heaven and Hell." In *The Complete Poetry and Prose of William Blake (Newly Revised Edition)*, edited by William V. Erdman, 33–45. New York: Anchor Books, 1988.

———. *Milton: A Poem in 2 Books*. In *The Complete Poetry and Prose of William Blake (Newly Revised Edition)*, edited by William V. Erdman, 95–143. New York: Anchor Books, 1988.

Bloch, Ernst. *Heritage of Our Times*. Translated by Neville Plaice and Stephen Plaice. Berkeley: University of California Press, 1990.

———. *The Principle of Hope, Volumes 1–3*. Translated by Neville Plaice, Stephen Plaice, and Paul Knight. Cambridge, MA: MIT Press, 1986.

Bloch, Ernst, Theodore Adorno, and Horst Kruger. "Something's Missing: A Discussion between Ernst Bloch and Theodor W. Adorno on the Contradictions of Utopian Longing." In *The Utopian Function of Art and Literature: Selected Essays*, translated by Jack Zipes and Frank Mecklenburg, 1–17. Cambridge, MA: MIT Press, 1989.

Blunt, Alison, and Gillian Rose, eds. *Writing Women and Space: Colonial and Postcolonial Geographies*. New York: Guilford Press, 1994.

Bose, Sugata. "Nation as Mother: Representations and Contestations of 'India' in Bengali Literature and Culture." In *Nationalism, Democracy, and Development: State and Politics in India*, edited by Sugata Bose and Ayesha Jalal, 50–75. New Delhi: Oxford University Press, 1997.

———. *Peasant Labour and Colonial Capital: Rural Bengal Since 1770*. Cambridge: Cambridge University Press, 1993.

Brantlinger, Patrick. *Rule of Darkness: British Literature and Imperialism, 1830–1914*. Ithaca: Cornell University Press, 1988.

Brennan, Timothy. "Subaltern Stakes." *New Left Review* 89 (2014): 67–87.

Brhad-aranyaka Upanishad. In *The Principal Upanishads*, edited and translated by S. Radhakrishnan. London: George Allen & Unwin Ltd., 1953.

Brooks, Peter. *The Melodramatic Imagination: Balzac, Henry James, Melodrama, and the Mode of Excess*. New Haven: Yale University Press, 1976.

Brown, Nicholas. *Utopian Generations: The Political Horizon of Twentieth-Century Literature*. Princeton, NJ: Princeton University Press, 2005.

Burke, Edmund. *A Philosophical Enquiry into the Origin of Our Ideas of the Sublime and Beautiful*. London: Routledge and Paul, 1958.

Buzzard, James. "The grand tour and after (1660–1840)." In *The Cambridge Companion to Travel Writing*, edited by Peter Hulme and Tim Youngs, 37–52. Cambridge: Cambridge University Press, 2002.

Carlyle, Thomas. "Occasional Discourse on the Negro Question." *Fraser's Magazine for Town and Country* XL, no. CCXL (December 1849): 670–678.

Casanova, Pasacale. *The World Republic of Letters*. Translated by M. B. De Bevoise. Cambridge, MA: Harvard University Press, 2004.

Chakrabarti, Hiren. "The Victoria Memorial." In *Calcutta: The Living City, Volume I: The Past*, edited by Sukanta Chaudhuri, 261–264. Calcutta: Oxford University Press, 1990.

Chakrabarty, Dipesh. *Provincializing Europe: Postcolonial Thought and Historical Difference*. Princeton, NJ: Princeton University Press, 2000.

Chakraborty, Chandrima. *Masculinity, Asceticism, Hinduism: Present and Past Imaginings of India*. Ranikhet: Permanent Black, 2011.

Chakravarty, Gautam. *The Indian Mutiny and the British Imagination*. Cambridge: Cambridge University Press, 2005.

Chatterji [Chatterjee], Bankim Chandra. *Anandamath* [আনন্দমঠ], edited by Brajendranath Bandyopadhyay and Sajonikanto Sen. Calcutta: Bangiya Sahitya Parishat, 1938.

———. *Anandamath, or, the Sacred Brotherhood*. Translated by Julius Lipner. New York: Oxford University Press, 2005.

———. *Dawn Over India*. Translated by Basanta Koomar Roy. New York: The Devin-Adair Company, 1941.

———. *Debi Chaudhurani, Or, the Wife Who Came Home*. Translated by Julius Lipner. New Delhi: Oxford University Press, 2010.

———. *Durgesa Nandini or, The Chieftain's Daughter: A Bengali Romance*. Translated by Charu Chandra Mookerjee. Calcutta: H. M. Mookerjee & Co., 1880.

Chatterji [Chatterjee], Bunkim [sic] Chandra. *Kopal-Kundala: A Tale of Bengal Life*. Translated by H. A. D. Phillips. London: Trübner & Co., *1885*.

Chatterji, Joya. *Bengal Divided: Hindu Communalism and Partition, 1932–1947*. New Delhi: Cambridge University Press, 2006.

———. *Krishna Kanta's Will*. Translated by Miriam S. Knight. London: T. Fischer Unwin, 1895.

———. *The Poison Tree: A Tale of Hindu Life in Bengal*. Translated by Miriam S. Knight. London: T. Fischer Unwin, 1884.

Chatterjee, Partha. *Nationalist Thought and the Colonial World: A Derivative Discourse?* Minneapolis, MN: University of Minnesota Press, 1986.

———. *The Black Hole of Empire: History of a Global Practice of Power*. Princeton, NJ: Princeton University Press, 2012.

———. *The Nation and its Fragments: Colonial and Postcolonial Histories*. Princeton, NJ: Princeton University Press, 1993.

Chattopadhyay, Geeta. *Bangla Swadeshi Gaan* [বাংলা স্বদেশী গান]. New Delhi: Delhi University, 1983.

Chattopadhyay, Swati. "Blurring Boundaries: The Limits of 'White Town' in Colonial Calcutta." *Journal of the Society of Architectural Historians* 59 (2000): 154–179.

———. "Cities and Peripheries." *Historical Research* 83, no. 222 (2010): 649–671.

———. *Representing Calcutta: Modernity, Nationalism, and the Colonial Uncanny*. London: Routledge, 2005.

Chaudhuri Rosinka. *The Literary Thing: History, Poetry, and the Making of a Modern Cultural Sphere*. New Delhi: Oxford University Press, 2014.

Chaudhuri, Supriya. "The Nation and Its Fictions: History and Allegory in Tagore's *Gora*." *South Asia: Journal of South Asian Studies* 35, no. 1 (2012): 97–117.

Chaudhury, Zahid. *Afterimage of Empire: Photography in Nineteenth-Century India*. Minneapolis, MN: University of Minnesota Press, 2012.

———. "Phantasmagoric Aesthetics: Colonial Violence and the Management of Perception." *Cultural Critique* 59 (2005): 63–119.

Chibber, Vivek. *Postcolonial Theory and the Specter of Capital*. London: Verso, 2012.

Chopra, Preeti. *A Joint Enterprise: Indian Elites and the Making of British Bombay*. Minneapolis, MN: University of Minnesota Press, 2011.

Chowdhury, Indira. *The Frail Hero and Virile History: Gender and the Politics of Culture in Colonial Bengal*. New Delhi: Oxford University Press, 1998.

Chunder, Bholanath. *The Travels of a Hindoo to Various Parts of Bengal and Upper India, Vol. II*. London: N. Trübner, 1869.

Clark, T. W., ed. *The Novel in India: Its Birth and Development*. Berkeley: University of California Press, 1970.

Conrad, Joseph. *Heart of Darkness*, edited by Robert Hampson. London: Penguin Books, 2007.

Cosgrove, Denis. "Prospect, Perspective and the Evolution of the Landscape Idea." *Transactions of the Institute of British Geographers* 10, no. 1 (1985): 45–62.

———. *Social Formation and Symbolic Landscape*. London: Croom Helm, 1984.

Curzon, George N. *British Government in India: The Story of the Viceroys and Government Houses, Volume I*. London: Cassell, 1925.

Das, Sisir Kumar. *The Artist in Chains: The Life of Bankimchandra Chatterji*. New Delhi: New Statesman Publishing, 1984.

Dasgupta, Subhoranjan. "Dystopia, Utopia, and Akhtaruzzaman Elias's Novel *Khowabnama*." In *The Politics of the (Im)Possible: Utopia and Dystopia Reconsidered*, edited by Barnita Bagchi, 80–94. New Delhi: Sage Publications, 2012.

Davis, Mike. *Late Victorian Holocausts: El Niño Famines and the Making of the Third World*. London: Verso, 2002.

Davis, Natalie Zemon. *Society and Culture in Early Modern France: Eight Essays*. Stanford: Stanford University Press, 1975.

Deb, Binay Krishna. *The Early History and Growth of Calcutta*. Calcutta: Romesh Chandra Ghose, 1905.

Derrida, Jacques. *Of Grammatology*. Translated by Gayatri Chakravorty Spivak. Baltimore: Johns Hopkins University Press, 2016.

Dev, Amiya. "Comparative Literature in India." *CLCWeb: Comparative Literature and Culture* 2, no. 4 (2000). https://doi.org/10.7771/1481-4374.1093

Dutton, Jacqueline. " 'Non-western Utopian Traditions." In *The Cambridge Companion to Utopian Literature*, edited by Gregory Claeys, 223–258. Cambridge: Cambridge University Press, 2010.

Dutton, Jacqueline, and Lyman Tower Sargent. "Introduction: Utopias from Other Cultural Traditions." *Utopian Studies* 24, no. 1 (2013): 2–5.

Eagleton, Terry. *Ideology, An Introduction*. London: Verso, 1991.

Elias, André J. P. " 'Vande Mataram!': Constructions of Gender and Music in Indian Nationalism." *Asian Music* 48, no. 2 (2017): 90–110.

Embree, Ainslee. "Frontiers into Boundaries: From the Traditional to the Modern State." In *Realm and Region in Traditional India, Volume 2*, edited by R. Fox, 255–270. Durham, NC: Duke University Press, 1977.

The Encyclopaedia Britannica, 11th ed., vol. 6. "Chatterji, Bankim Chandra." Chicago: Encyclopedia Britannica, 1910.

Fabian, Johannes. *Time and the Other: How Anthropology Makes its Object*. New York: Columbia University Press, 1983.

Fanon, Frantz. *The Wretched of the Earth*. Translated by Constance Farrington. New York: Grove Press, 1963.

Fay, Eliza. *Original Letters from India (1779–1815)*. Edited by E. M. Forster. London: Chatto & Windus, 1986.

Flint, Kate. *The Victorians and the Visual Imagination*. Cambridge: Cambridge University Press, 2000.

Forster, E. M. *A Passage to India*. Harmondsworth: Penguin Books, 1924.

Foucault, Michel. *The History of Sexuality, Volume I: An Introduction*. Translated by Robert Hurley. New York: Pantheon Books, 1978.

Frye, Northrop. *The Anatomy of Criticism: Four Essays*. Princeton, NJ: Princeton University Press, 1971.

———. *The Secular Scripture: A Study of the Structure of Romance*. Boston: Harvard University Press, 1976.

Fukuyama, Francis. *The End of History and the Last Man*. New York: Free Press, 1992.

Gajarawala, Toral Jatin. *Untouchable Fictions: Literary Realism and the Crisis of Caste*. New York: Fordham University Press, 2013.

Gandhi, Mohandas K. "Village Swaraj." *Harijan* 9, no. 28 (1942): 238.

———. "The Great Sentinel." In *The Mahatma and The Poet: Letters and Debates between Gandhi and Tagore 1915–1941*, edited by Sabyasachi Bhattacharya, 92–97. New Delhi: National Book Trust, 1997.

Ghalib, Mirza. "Conversation with His Guiding Angel." In *The Shadow of a Bird in Flight*, edited and translated by Shamsur Rahman Faruqi, 44–49. New Delhi: Rupa, 1994.

———. *Divan-i Ghalib Dihlavi*, edited by Muhsin Kiyani. Tehran: Rawzanah, 1997.

Gibney, Robert D. *My Escape from the Mutinies in Oudh*. London: Richard Bentley, 1858.

Gidwani, Vinay. " 'Waste' and the Permanent Settlement in Bengal." *Economic and Political Weekly* 27, no. 4 (1992): PE39–PE46.

Gilpin, William. *Three Essays: On Picturesque Beauty: On Picturesque Travel: And on Sketching Landscape: To which is Added a poem, on Landscape Painting*. London: Printed for R. Blamire, 1792.

Girard, René, *Deceit, Desire, and the Novel: Self and Other in Literary Structure*. Baltimore: Johns Hopkins Press, 1965.

Gledhill, Christine. "The Melodramatic Field: An Investigation." In *Home is Where the Heart Is*, edited by Christine Gledhill, 5–39. London: British Film Institute, 1987.

Gombrich, Ernst H. "Cartoonists' Armoury." In *Meditations on a Hobby Horse: And Other Essays on the Theory of Art*, 127–142. London: Phaidon, 1971.

Gopal, Priyamvada. *Literary Radicalism in India: Gender, Nation and the Transition to Independence*. London: Routledge, 2005.

———. "This Is a Watershed Moment for India. It Must Choose Freedom Over Intolerance." *The Guardian*, February 17, 2016.

Gopal, Ram. *A Political History of Indian Muslims, 1858–1947*. New Delhi: Criterion Publications, 1988.

Goswami, Manu. *Producing India: From Colonial Economy to National Space*. Chicago: University of Chicago Press, 2004.

Govind, Nikhil. *Between Love and Freedom: The Revolutionary in the Hindi Novel*. New Delhi: Routledge, 2014.

Graham, James. *Land and Nationalism in Fictions from Southern Africa*. New York: Routledge, 2009.

Gramsci, Antonio. "Notes for an Introduction and an Approach to the Study of Philosophy and the History of Culture." In *The Gramsci Reader: Selected Writings 1916–1935*, edited by David Forgacs, 324–343. New York: New York University Press, 2000.

Gupta, Narayani. *Delhi between Two Empires, 1803–1931: Society, Government and Urban Growth*. New Delhi: Oxford University Press, 1981.

Guha, Ranajit. *A Rule of Property for Bengal: An Essay on the Idea of Permanent Settlement*. Paris: Mouton, 1963.

Gutiérrez, Gustavo. *A Theology of Liberation: History, Politics, and Salvation*. Maryknoll: Orbis Books, 1973.

Harris, G. *A Lady's Diary of the Siege of Lucknow: Written for the Perusal of Friends at Home*. London: J. Murray, 1858.

Heber, Reginald. "Before A Collection for the Society for the Propagation of the Gospel." In *The Poetical Works of Reginald Heber*, 86–88. Boston: Little Brown & Co., 1853.

Heehs, Peter. *The Bomb in Bengal: The Rise of Revolutionary Terrorism in India, 1900–1910*. New Delhi: Oxford University Press, 1993.

Heidegger, Martin. "The Age of the World Picture." In *The Question Concerning Technology and Other Essays*, translated by William Lovitt, 115–154. New York: Harper Perennial, 1977.

Hickey, William. *Memoirs of William Hickey, Vol. 2*. Edited by Alfred Spencer. London: Hurst & Blackett, 1916.

Hill, Christopher. *Milton and the English Revolution*. New York: Viking Press, 1978.

———. *The World Turned Upside Down: Radical Ideas During the English Revolution*. Harmondsworth: Penguin Books, 1975.

Hodder, Edwin. *Cities of the World: Their Origin, Progress, and Present Aspect*. New York, NY: Cassell & Co., 1886.

Hopkirk, Peter. *Quest for Kim: In Search of Kipling's Great Game*. Ann Arbor: University of Michigan Press, 1997.

———. *The Great Game: On Secret Service in High Asia*. London: Murray, 1990.

———. *Trespassers on the Roof of the World: The Race for Lhasa*. Oxford: Oxford University Press, 1983.

Hossain, Rokeya Sakhawat. *Padmarag* [পদ্মরাগ]. In *Begum Rokeya Rachanabali* [বেগম রোকেয়া রচনাবলী], edited by Mostafa Mir, 261–368. Dhaka: Bornayon, 2000.

———. *Padmarag*. In *Sultana's Dream and Padmarag: Two Feminist Utopias*, edited by Barnita Bagchi, 15–193. New Delhi: Penguin Books, 2005.

Ingold, Tim. *The Perception of the Environment: Essays on Livelihood, Dwelling and Skill*. London: Routledge, 2002.

Islam, Sirajul. *The Permanent Settlement in Bengal: A Study of Its Operation 1790–1819*. Dhaka: University Press, 1979.

Jadhav, Rajendra. "Thousands of farmers march to India's Mumbai to demand help." *Reuters US*, March 12, 2018.

Jameson, Fredric. *The Antinomies of Realism*. London: Verso, 2013.

———. *Archaeologies of the Future: The Desire Called Utopia and Other Science Fictions*. London: Verso, 2005.

———. *Late Marxism: Adorno, or the Persistence of the Dialectic*. London: Verso, 1990.

———. *The Political Unconscious: Narrative as a Socially Symbolic Act*. Ithaca: Cornell University Press, 1982.

———. *A Singular Modernity: Essay on the Ontology of the Present*. London: Verso, 2002.

———. *Signatures of the Visible*. New York: Routledge, 1990.

———. "Utopia as Method, or the Uses of the Future." In *Utopia/Dystopia: Conditions of Historical Possibility*, edited by M. D. Gordin, H. Tilley, and G. Prakash, 21–44. Princeton, NJ: Princeton University Press, 2010.

Johnson, Alan G. *Out of Bounds: Anglo-Indian Literature and the Geography of Displacement*. Honolulu: University of Hawaii Press, 2011.

Jones, Gareth S. *Outcast London: A Study in the Relationship between Classes in Victorian Society*. Oxford: Clarendon Press, 1971.

Joshi, Priya. *In Another Country: Colonialism, Culture, and the English Novel in India*. New York: Columbia University Press, 2002.

———. "Globalizing Victorian Studies." *The Yearbook of English Studies* 41, no. 2 (2011): 20–40.

Kaiwar, Vasant. *The Postcolonial Orient: The Politics of Difference and the Project of Provincialising Europe*. Leiden: Brill, 2014.

Kant, Immanuel. *Critique of Judgment*. Translated by J. C. Meredith. Oxford: Clarendon Press, 1952.

Kapila, Shuchi. *Educating Seeta: The Anglo-Indian Family Romance and the Poetics of Indirect Rule*. Columbus: Ohio State University Press, 2010.

Karatani, Kōjin. *Origins of Modern Japanese Literature*. Durham, NC: Duke University Press, 1993.

Kaul, Suvir. *Poems of Nation, Anthems of Empire: English Verse in the Long Eighteenth Century*. Charlottesville: University Press of Virginia, 2000.

Kaviraj, Sudipta. *The Invention of Private Life: Literature and Ideas*. New York: Columbia University Press, 2015.

———. "Making of a Language of Patriotism in Modern Bengali." In *Nationalism and Ethnosymbolism: History, Culture and Ethnicity in the Formation of Nations*, edited by Athena S. Leoussi and Steven Grosby, 248–264. Edinburgh: Edinburgh University Press, 2007.

———. *The Unhappy Consciousness: Bankimchandra Chattopadhyay and the Formation of Nationalist Discourse in India*. New Delhi: Oxford University Press, 1995.

Keshavamurthy, Kiran. *Beyond Desire: Sexuality in the Modern Tamil Novel*. New Delhi: Oxford University Press, 2016.

Keshavmurthy, Prashant. *Persian Authorship and Canonicity in Late Mughal Delhi: Building an Ark*. London: Routledge, 2016.

Kiernan, Victor G. *The Lords of Human Kind: Black Man, Yellow Man, and White Man in an Age of Empire*. Boston: Little Brown, 1969.

King, Anthony D. *Colonial Urban Development: Culture, Social Power and Environment*. London: Routledge and Kegan Paul, 1976.

Kipling, Rudyard. *The City of Dreadful Night and Other Places*. London: Sampson Low, Marston & Co., 1899.

———. *Kim*, edited by Maire ni Fhlathuin. Peterborough: Broadview Press, 2005.

———. "Lispeth." In *Plain Tales from the Hills*, 1–7. New York: Doubleday & McClure Company, 1899.

Klein, Ira. "Death in India, 1871–1921." *The Journal of Asian Studies* 32, no. 4 (1973): 639–659.

Kopf, David. *The Brahmo Samaj and the Shaping of the Modern Indian Mind*. Princeton, NJ: Princeton University Press, 1979.

Koven, Seth. *Slumming: Sexual and Social Politics in Victorian London*. Princeton, NJ: Princeton University Press, 2004.

Krishna, Shankaran. "Cartographic Anxiety: Mapping the Body Politic in India." In *Challenging Boundaries: Global Flows, Territorial Identities*, edited by M. Shapiro and H. Alker, 193–214. Minneapolis, MN: University of Minnesota Press, 1996.

Kumar, Udaya. *Writing the First Person: Literature, History, and Autobiography in Modern Kerala*. Ranikhet: Permanent Black, 2016.

Lazarus, Neil. *Nationalism and Cultural Practice in the Postcolonial World*. Cambridge: Cambridge University Press, 1999.

———. *The Postcolonial Unconscious*. Cambridge: Cambridge University Press, 2011.

Lefebvre, Henri. *The Production of Space*. Translated by Donald Nicholson-Smith. Oxford: Blackwell, 1991.

———. *Writings on Cities*. Translated by Eleonore Kofman and Elizabeth Lebas. Malden: Blackwell Publishing, 1996.

Levitas, Ruth. *The Concept of Utopia*. Oxford: Peter Lang, 1990.

———. *Utopia as Method: The Imaginary Reconstitution of Society*. New York: Palgrave Macmillan, 2013.

Lipner, Julius J. " 'Icon and Mother': An Inquiry into India's National Song." *The Journal of Hindu Studies* 1, no. 1–2 (2008): 26–48.

———. *Introduction to Anandamath or The Sacred Brotherhood*, by Bankim Chatterji, translated by Julius J. Lipner, 3–124. New York: Oxford University Press, 2005.

Lukacs, Georg. "Reification and the Consciousness of the Proletariat." In *History and Class Consciousness: Studies in Marxist Dialectics*, translated by Rodney Livingstone, 83–222. Cambridge, MA: MIT Press, 1971.

Majeed, Javed. *Ungoverned Imaginings: James Mill's the History of British India and Orientalism*. Oxford: Clarendon Press, 1992.

Majumder, Atreyee. *Time, Space and Capital in India: Longing and Belonging in an Urban-Industrial Hinterland*. London: Routledge, 2018.

Majumder, Auritro. "Can Bengali Literature be Postcolonial?" *Comparative Literature Studies* 53, no. 2 (2016): 417–425.

Malamoud, Charles. "Bricks and Words: Observations on the Bodies of the Gods in Vedic India." In *Cooking the World: Ritual and Thought in Ancient India*, translated by David White, 207–225. New Delhi: Oxford University Press, 1996.

Marin, Louis. *Utopics: The Semiological Play of Textual Spaces*. Translated by Robert A. Vollrath. Amherst: Humanity Books, 1984.

Marx, Karl. *Capital: A Critique of Political Economy, Volume I, The Process of Capitalist Production*. Translated by Samuel Moore and Edward Aveling. London: Lawrence & Wishart Ltd., 1967.

Mbembe, Achille. *On the Postcolony*. Berkeley: University of California Press, 2001.

McClintock, Anne. *Imperial Leather: Race, Gender, and Sexuality in the Colonial Contest*. New York: Routledge, 1995.

McKean, Lise. "Bharat Mata: Mother India and Her Militant Matriots." In *Devī: Goddesses of India*, edited by John Stratton Hawley and Donna Marie Wulff, 250–280. Berkeley: University of California Press, 1996).

Mearns, Andrew. *The Bitter Cry of Outcast London*. London: James Clarke & Co., 1883.

Mee, Jon. *Dangerous Enthusiasm: William Blake and the Culture of Radicalism in the 1790s*. Oxford: Clarendon Press, 1994.

Mehta, Jaya. " 'Some Imaginary "Real" Thing': Racial Purity, the Mutiny, and the Nation in Tagore's *Gora* and Kipling's *Kim*." In *Rabindranath Tagore: Universality and Tradition*, edited by Patrick Colm Hogan and Lalita Pandit, 199–212. Madison: Farleigh Dickinson University Press, 2003.

Meisami, Julie Scott. *Medieval Persian Court Poetry*. Princeton, NJ: Princeton University Press, 1987.

Metcalf, Thomas. *Ideologies of the Raj*. Cambridge: Cambridge University Press, 1998.

———. *An Imperial Vision: Indian Architecture and Britain's Raj*. Oxford: Oxford University Press, 2002.

Milton, John. *Paradise Lost*, edited by Alastair Fowler. Harlow: Longman, 2007.

Mir, Farina. *The Social Space of Language: Vernacular Culture in British Colonial Punjab*. Berkeley: University of California Press, 2010.

Mishra, Atul. "Boundaries and Territoriality in South Asia: From Historical Comparisons to Theoretical Considerations." *International Studies* 45 (2008): 105–132.

Mitchell, Don. *The Lie of the Land: Migrant Workers and the California Landscape*. Minneapolis, MN: University of Minnesota Press, 1996.

Mitchell, E. *Thacker's Guidebook to Calcutta: Its High-ways & By-paths, With a Chapter on the Government of India*. Calcutta: Thacker, Spink & Co., 1890.

Mitchell, W. J. T. "Imperial Landscape." In *Landscape and Power*, edited by W. J. T. Mitchell, 5–30. Chicago: University of Chicago Press, 2002.

———. *Introduction to Landscape and Power*, edited by W. J. T. Mitchell, 1–4. Chicago: University of Chicago Press, 2002.

Mohan, Anupama. *Utopia and the Village in South Asian Literatures*. Basingstoke: Palgrave Macmillan, 2012.

Moretti, Franco. "Conjectures on World Literature." *New Left Review* 1 (2000): 54–68.

Morton, A. L. *The Everlasting Gospel: A Study in the Sources of William Blake*. New York: Haskell House, 1966.

Mukherjee, Meenakshi. *Realism and Reality: The Novel and Society in India*. New Delhi: Oxford University Press, 1985.

Mukherjee, Janam. *Hungry Bengal: War, Famine and the End of Empire*. New York: Oxford University Press, 2015.

Mukherjee, Soumyendra Nath. *Calcutta: Essays in Urban History*. Calcutta: Subarnarekha, 1993.

Mukherjee, Upamanyu Pablo. "Introduction: Victorian World Literatures." *The Yearbook of English Studies* 41, no. 2 (2011): 1–19.

———. *Natural Disasters and Victorian Empire: Famines, Fevers and the Literary Cultures of South Asia*. Basingstoke: Palgrave Macmillan, 2013.

Nair, Parameswaran Thankappan. *Calcutta in the 17th Century*. Calcutta: Firma KLM, 1986.

———. "Civic and Public Services in Old Calcutta." In *Calcutta: The Living City, Volume I: The Past*, edited by Sukanta Chaudhuri, 224–245. Calcutta: Oxford University Press, 1990.

———. "The Growth and Development of Old Calcutta." In *Calcutta: The Living City, Volume I: The Past*, edited by Sukanta Chaudhuri, 10–23. Calcutta: Oxford University Press, 1990.

Nandy, Ashish. "The Illegitimacy of Nationalism: Rabindranath Tagore and the Politics of the Self." In *Bonfire of Creeds*, 153–233. New Delhi: Oxford University Press, 2004.

Nehru, Jawaharlal. "The Appointed Day." In *Jawaharlal Nehru's Speeches, Volume One, September 1946 – May 1949*, 27–28. New Delhi: Publications Division, Government of India.

———. *The Discovery of India*. New Delhi: Oxford University Press, 1985.

———. "Tryst with Destiny." In *Jawaharlal Nehru's Speeches, Volume One, September 1946 – May 1949*, 25–26. New Delhi: Publications Division, Government of India.

Ngũgĩ wa Thiong'o. *Globalectics: Theory and the Politics of Knowing*. New York: Columbia University Press, 2012.

Noorani, A. G. "Vande Mataram: A Historical Lesson." *Economic and Political Weekly* 8, no. 23 (1973): 1039–1043.

Olwig, Kenneth. *Landscape, Nature, and the Body Politic: From Britain's Renaissance to America's New World*. Madison: University of Wisconsin Press, 2002.

Ojha, Krittibas. *Shochitro Shoptokando Ramayon* [সচিত্র সপ্তকান্ড রামায়ন]. Edited by Ramananda Chatterjee. Calcutta: Probashi Press, 1926.

Ong, Walter J. *Orality and Literacy: The Technologizing of the Word*. London: Routledge, 2002.

Orsini, Francesca. "Indian in the Mirror of World Fiction." *New Left Review* 13 (2002): 75–88.

Otter, Chris. *The Victorian Eye: A Political History of Light and Vision in Britain, 1800–1910*. Chicago: University of Chicago Press, 2008.

Pandit, Lalita. "Caste, Race, and Nation: History and Dialectic in Rabindranath Tagore's *Gora*." In *Literary India: Comparative Studies in Aesthetics, Colonialism, and Culture*, edited by Patrick Colm Hogan and Lalita Pandit, 207–233. Albany: State University of New York Press, 1995.

Parks, Fanny. *Wanderings of a Pilgrim in Search of the Picturesque, Vol. II*. Edited by Esther Chawner. Karachi: Oxford University Press, 1975.

Pinney, Christopher. *Camera Indica: The Social Life of Indian Photographs*. London: Reaktion Books, 1997.

Plowden, W. Chichele. *The Indian Empire Census of 1881 Statistics of Population Vol. II*. Calcutta: Superintendent of Government Printing, 1883.

Prasad, Rajendra. "Volume 12. 24th January, 1950" [12.167.2]. Constituent Assembly of India Debates (Proceedings) – Volume XII. http://cadindia.clpr.org.in/constitution_assembly_debates/volume/12/1950-01-24 (accessed November 15, 2018).

Pritchett, Frances W. *Nets of Awareness: Urdu Poetry and Its Critics*. Berkeley: University of California Press, 1994.

Radović, Stanka. *Locating the Destitute: Space and Identity in Caribbean Fiction*. Charlottesville: University of Virginia Press, 2014.

Rainey, H. J. *A Historical and Topographical Sketch of Calcutta*. Calcutta: "Englishman" Press, 1876.

Raj, Kapil. "The Historical Anatomy of a Contact Zone: Calcutta in the Eighteenth Century." *The Indian Economic and Social History Review* 48, no.1 (2011): 154–179.

Ramaswamy, Sumathi. *The Goddess and the Nation: Mapping Mother India*. Durham, NC: Duke University Press, 2010.

Ramazani, Jahan. *A Transnational Poetics*. Chicago: Chicago University Press, 2009.

Ray, Bharati. *Early Feminists of Colonial India: Sarala Devi Chaudhurani and Rokeya Sakhawat Hossain*. New Delhi: Oxford University Press, 2002.

Ray, Dwijendralal. "Bharotborsho" [ভারতবর্ষ]. In *Gaan* [গান], 3–4. Calcutta: Bengal Medical Library, 1915.

Ray, Sangeeta. *En-Gendering India: Woman and Nation in Colonial and Postcolonial Narratives*. Durham, NC: Duke University Press, 2000.

Richards, Thomas. *Imperial Archive: Knowledge and Fantasy of Empire*. London: Verso, 1993.

Rodrigues, Hillary. *Ritual Worship of the Great Goddess: The Liturgy of the Durga Puja with Interpretations*. Albany: State University of New York Press, 2003.

Rousseau, Jean-Jacques. *The Social Contract and Other Later Political Writings, Book IV*. Edited and Translated by V. Gourevitch. Cambridge: Cambridge University Press, 1997.

Roy, Modhumita. "Utopia in the Subjunctive Mood: Bessie Head's *When Rain Clouds Gather*." In *The Politics of the (Im)Possible: Utopia and Dystopia Reconsidered*, edited by Barnita Bagchi, 179–193. New Delhi: Sage Publications, 2012.

Russell, Ralph and Khurshidul Islam. *Ghalib: Life and Letters*. New Delhi: Oxford University Press, 1994.

Said, Edward. *Culture and Imperialism*. New York: Vintage Books, 1994.

———. *Orientalism*. New York: Vintage Books, 1979.

———. "Secular Criticism." In *The Text, the World and the Critic*, 1–30. Boston: Harvard University Press, 1984.

Sahota, G. S. *Late Colonial Sublime: Neo-Epics and the End of Romanticism*. Evanston: Northwestern University Press, 2018.

San Juan, E. "Postcolonialism and the Problematic of Uneven Development." In *Marxism, Modernity and Postcolonial Studies*, edited by Crystal Bartolovich and Neil Lazarus, 221–239. Cambridge: Cambridge University Press, 2002.

Sargent, Lyman T. "Colonial and Postcolonial Utopias." In *The Cambridge Companion to Utopian Literature*, edited by Gregory Claeys, 200–222. Cambridge: Cambridge University Press, 2010.

———. "The Three Faces of Utopianism Revisited." *Utopian Studies* 5, no. 1 (1994): 1–37.

Sarkar, Sumit. *The Swadeshi Movement in Bengal, 1903–1908*. New Delhi: People's Publishing House, 1973.

———. *Writing Social History*. New Delhi: Oxford University Press, 1997.

Sarkar, Tanika. "Birth of a Goddess: 'Vande Mataram,' *Anandamath*, and Hindu Nationhood." *Economic and Political Weekly* 41, no. 37 (2006): 3959–3969.

———. *Hindu Wife, Hindu Nation, Community, Religion, and Cultural Nationalism*. Bloomington: Indiana University Press, 2001.

Sartori, Andrew. *Bengal in Global Concept History: Culturalism in the Age of Capital*. Chicago: University of Chicago Press, 2008.

———. *Bengal in Global Concept History: Culturalism in the Age of Capital*. Chicago: University of Chicago Press, 2008.

———. *Liberalism in Empire: An Alternative History*. Berkley: University of California Press, 2014.

Savarkar, Vinayak Damodar. *Hindutva: Who Is a Hindu?* Bombay: Veer Savarkar Prakashan, 1969.

Schievelbusch, Wolfgang. *Disenchanted Night: The Industrialization of Light in the Nineteenth Century*. Berkeley: University of California Press, 1995.

Schwarz, Henry. "Sexing the Pundits: Gender, Romance, and Realism in the Cultural Politics of Colonial Bengal." In *Reading the Shape of the World: Toward an International Cultural Studies*, edited by Henry Schwarz and Richard Dienst, 224–258. Boulder: Westview Press, 1996.

Sen, Amartya. *Poverty and Famines: An Essay on Entitlement and Deprivation*. Oxford: Oxford University Press, 1983.

Sen, Prabodh Chandra. *India's National Anthem*. Calcutta: Vishva-Bharati, 1972.

Sen, Samita. "A Parliament of Women: Dystopia in Nineteenth-Century Bengali Imagination." In *The Politics of the (Im)Possible: Utopia and Dystopia Reconsidered*, edited by Barnita Bagchi, 123–146. New Delhi: Sage Publications, 2012.

Sen, Sudipta. *Empire of Free Trade: The East India Company and the Making of the Colonial Marketplace*. Philadelphia: University of Pennsylvania Press, 1998.

Shandilya, Krupa. *Intimate Relations: Social Reform and the Late Nineteenth Century South Asian Novel*. Evanston: Northwestern University Press, 2017.

Shankar, Subramanian. *Flesh and Fish Blood: Postcolonialism, Translation, and the Vernacular*. Berkeley: University of California Press, 2012.

Sharma, Sunil. "The City of Beauties in Indo-Persian Poetic Landscape." *Comparative Studies of South Asia, Africa and the Middle East* 24, no. 2 (2004): 73–81.

Simms, Frederick Walter. *Report of the Survey of Calcutta*. Calcutta: Military Orphan Press, 1851.

Sinha, Kaliprasanna. *The Observant Owl: Hootum's Vignettes of Nineteenth-Century Calcutta*. Translated by Swarup Roy. Ranikhet: Black Kite, 2008.

Sinha, Mrinalini. *Colonial Masculinity: The 'Manly Englishman' and the 'Effeminate Bengali' in the Late Nineteenth Century*. Manchester: Manchester University Press, 1995.

Sinha, Pradip. *Calcutta in Urban History*. Calcutta: Firma KLM, 1978.

Smith, David. *Report on the Drainage and Conservancy of Calcutta*. Calcutta: Bengal Secretariat Press, 1869.

Smith, Neil. *Uneven Development: Nature, Capital, and the Production of Space*. Athens: University of Georgia Press, 1990.

Spivak, Gayatri Chakravorty. "Can the Subaltern Speak?" In *Marxism and the Interpretation of Culture*, edited by Cary Nelson and Lawrence Grossberg, 271–313. Urbana: University of Chicago Press, 1988.

———. *Death of a Discipline*. New York: Columbia University Press, 2003.

———. "Foreword to the Routledge Classics Edition." *Outside in the Teaching Machine*, xi–xv. New York: Routledge, 2009.

———. "Resident Alien." In *An Aesthetic Education in the Era of Globalization*, 301–315. Cambridge: Harvard University Press, 2012.

Spurr, David. *The Rhetoric of Empire: Colonial Discourse in Journalism, Travel Writing, and Imperial Administration*. Durham, NC: Duke University Press, 1993.

Stallybrass, Peter and Allon White. *The Politics and Poetics of Transgression*. Ithaca: Cornell University Press, 1986.

Stavorinus, Johan. *Voyages to the East Indies, Vol. I*. London: G. G. & J. Robinson, 1798.

Stokes, Eric and C. A. Bayly. *The Peasant Armed: The Indian Revolt of 1857*. Oxford: Clarendon Press, 1986.

Strachey, John. *India*. London: Kegan Paul, Trench & Co., 1888.

Suhrud, Tridip. *Writing Life: Three Gujarati Thinkers*. New Delhi: Orient Blackswan, 2009.

Tagore, Rabindranath. "amar shonar Bangla" [আমার সোনার বাংলা]. In *Akhando Gitobitan* [অখন্ড গীতবিতান], 243. Calcutta: Vishwabharati, 1993.

———. "Bastob" [বাস্তব]. In *Sahityer Pothe* [সাহিত্যের পথে], Rabindra Rachanabali [রবীন্দ্র রচনাবলী], vol. 23, 361–368. Calcutta: Vishwabharati, 1958.

———. "Bharot-tirtho" [ভারততীর্থ]. In *Akhando Gitobitan* [অখন্ড গীতবিতান], 251. Calcutta: Vishwabharati, 1993.

———. "Bongomata" [বঙ্গমাতা]. In *Chaitali* [চৈতালি], 58. Calcutta: Vishwabharti, 1896.

———. "Chitto jetha bhoyshunyo" [চিত্ত যেথা ভয়শূন্য]. In *Noibedyo* [নৈবেদ্য], 83. Calcutta: Vishwabharati, 1901.

———. *Gora* [গোরা]. Calcutta: Vishwabharati, 2010.

———. *Home and the World*. Translated by Surendranath Tagore. London: Macmillan & Co., 1919.

———. "Jonoganomono-odhinayako" [জনগনমনঅধিনায়ক]. In *Akhando Gitobitan* [অখন্ড গীতবিতান], 249–250. Calcutta: Vishwabharati, 1993.

———. "Letter 314. To Subhas Chandra Bose." In *Selected letters of Rabindranath Tagore*, edited by Krishna Dutta and Andrew Robinson, 486–488. Cambridge: Cambridge University Press, 1997.

———. "My Bengal of Gold." In *The Essential Tagore*, edited by Fakhrul Alam and Radha Chakravarty, translated by Fakhrul Alam, 330–331. Cambridge, MA: The Belknap Press of Harvard University Press, 2011.

———. *My Boyhood Days*. Translated by Marjorie Sykes. Calcutta: Viswabharati, 1945.

———. "Pilgrimage to India." In *Rabindranath Tagore: Selected Poems*, edited and translated by Sukanta Chaudhuri, 200–202. New Delhi: Oxford University Press, 2004.

Tally Jr., Robert, ed. *Geocritical Explorations: Space, Place and Mapping in Literary and Cultural Studies*. New York: Palgrave Macmillan, 2011.

———. *Utopia in the Age of Globalization: Space, Representation, and the World-System*. Basingstoke: Palgrave Macmillan, 2013.

Taylor, Philip Meadows. *Seeta*. London: C. Kegan Paul & Co., 1880.

Telscher, Kate. "India/ Calcutta: City of Palaces and Dreadful Night." In *The Cambridge Companion to Travel Writing*, edited by Peter Hulme and Tim Youngs, 191–206. Cambridge: Cambridge University Press, 2002.

Teverson, Andrew, and Sara Upstone, eds. *Postcolonial Spaces: The Politics of Place in Contemporary Culture*. Basingstoke: Palgrave Macmillan, 2011.

Thieme, John. *Postcolonial Literary Geographies: Out of Place*. London: Palgrave Macmillan, 2016.

Thomas, Keith. *Religion and the Decline of Magic: Studies in Popular Beliefs in Sixteenth and Seventeenth Century England*. Harmondsworth: Penguin Books, 1982.

Thompson, Edward P. "Time, Work-discipline, and Industrial Capitalism." *Past and Present* 38 (1967): 56–97.

———. *Witness Against the Beast: William Blake and the Moral Law*. New York: New Press, 1993.

Thompson, Mowbray. *Story of Cawnpore*. London: Richard Bentley, 1859.

Thomson, James. "Rule Britannia." In *The Oxford Book of Eighteenth Century Verse*, edited by David Nichol Smith, 252–253. London: Oxford University Press, 1951.

Trevelyan, George. *Cawnpore*. London: Macmillan and Co., 1866.

Upstone, Sara. *Spatial Politics in the Postcolonial Novel*. Farnham: Ashgate, 2009.

Varma, Rashmi. *The Postcolonial City and its Subjects: London, Nairobi, Bombay*. New York: Routledge, 2012.

Virgil. *Aeneid*. Translated by Frederick Ahl. Oxford: Oxford University Press, 2007.

Visaria, Leela, and Pravin Visaria. "Population, 1757–1947." In *Cambridge Economic History of India, Volume 2: c.1757–1970*, edited by Dharma Kumar and Meghnad Desai, 463–552. Cambridge: Cambridge University Press, 1983.

Walkowitz, Judith R. *Prostitution and Victorian Society: Women, Class, and the State*. Cambridge: Cambridge University Press, 1980.

Wallerstein, Immanuel M. *The Capitalist World-Economy: Essays*. Cambridge: Cambridge University Press, 1979.

Warwick Research Collective (WReC). *Combined and Uneven Development: Towards a New Theory of World-Literature*. Liverpool: Liverpool University Press, 2015.

Waterfield, Henry. *Memorandum on the Census of British India 1871–72*. London: Her Majesty's Stationary Office, 1875.

Watson, John Forbes, and John W. Kaye, eds. "Preface." In *The People of India: A Series of Photographic Illustrations, with Descriptive Letterpress, of the Races and Tribes of Hindustan: Originally Prepared Under the Authority of the Government of India and Reproduced by Order of the Secretary of State for India in Council, Volume One*. London: India Museum, 1868.

Wegner, Philip E. *Imaginary Communities: Utopia, Nation, and the Spatial Histories of Modernity*. Berkeley: University of California Press, 2002.

———. " 'Life as He Would Have It': The Invention of India in Kipling's *Kim*." *Cultural Critique* (1993): 129–159.

Wenzel, Jennifer. "Decolonization." In *A Companion to Critical and Cultural Theory*, edited by Imre Szeman, Sarah Blacker, and Justin Sully, 449–464. Hoboken: Wiley Blackwell, 2017.

Williams, Linda. "Melodrama Revised." In *Refiguring American Film Genres: History and Theory*, edited by Nick Browne, 42–88. Berkeley: University of California Press, 1998.

Williams, Raymond. *The Country and the City*. New York: Oxford University Press, 1973.

———. *Marxism and Literature*. Oxford: Oxford University Press, 1977.

Wilson, Angus. *The Strange Ride of Rudyard Kipling: His Life and Works*. London: Secker and Warburg, 1977.

Winichakul, Thongchai. *Siam Mapped: A History of the Geo-Body of a Nation*. Honolulu: University of Hawaii Press, 1994.

Zacharias, Robert. "Space and the Postcolonial Novel." In *The Cambridge Companion to the Postcolonial Novel*, edited by Ato Quayson, 208–229. New York: Cambridge University Press, 2016.

Zvelebil, K. V. *Introduction to The Lord of the Meeting Rivers: Devotional Poems of Basavanna*, 1–4. New Delhi: Motilal Banarsidass, 1984.

Index

1857 Uprising 18–21, 23, 24, 29, 30, 32, 34, 88–89, 120–121, 131, 137

abstract space *see* Lefebvre, Henri
Adorno, Theodor 5, 90, 148
adventure time *see* Bakhtin, Mikhail
affective impulse *see* Jameson, Frederic
affiliation *see* Said, Edward
age of the world-picture *see* Heidegger, Martin
Ambedkar, BR 134
Anderson, Benedict 4
Atkinson, James 10, 51, 57, 58; *City of Palaces* 55–56

Bagchi, Barnita 15n40, 117, 140, 143
Baig, Mirza Asadullah *see* Ghalib, Mirza
Bakhtin, Mikhail 36, 39; adventure time 36; chronotope 39–41; heteroglossia 11, 107, 117, 128–137
Bandyopadhyay, Bhabanicharan 51, 68, 69, 75, 129; *Kalikata Kamalalay* 58–59
Basu, Subho 14n24, 80n64, 81n97, 113n54
Battle of Plassey 33, 52
Bengal 10, 19, 21, 24, 33–34, 35, 36, 37, 38, 39–42, 43, 44, 45, 52–54, 55, 60, 61, 62, 67, 73, 77, 82, 83, 94, 88, 91, 92–99, 101, 117, 128, 129, 133–136, 138, 141–142, 143
Bengali (language and literature) 6, 7, 9, 11, 19, 32, 33, 45, 49n77, 58, 59, 85, 92, 97, 98, 99–100, 106, 113n63, 113n64, 116, 128, 129, 132, 137, 140, 142, 146n82, 154–156
Bengalis 34, 43, 44, 45, 51, 54, 67, 68, 71, 72, 73, 93–96, 124–125, 129, 130, 131, 133, 134, 135, 137, 138, 141–142, 143; bhadralok 131, 135
Benjamin, Walter 5
Bhadralok *see* Bengalis
Bhatia, B.M. 94
Bhattacharya, Tithi 49n84
Blake, William 85, 108; "mind forg'd manacles" 149
Bloch, Ernst 5, 7, 19, 114n82, 148; Not Yet 5, 7, 37, 148, 149; Not-Yet-Become 5; Not-Yet-Conscious 5; simultaneity of the non-simultaneous 119
Bongodorshon 10, 32, 33, 44, 82, 94
border *see* space
Bose, Subhash Chandra 83
Bose, Sugata 4, 95
boundary *see* space
Brahmo, Brahmo Sabha, Brahmo Samaj 107, 117, 129, 130, 138
Buddha, Buddhism 28, 35, 101, 102, 118, 119, 121, 123, 126, 127, 154

Calcutta 10, 24, 33, 51–77, 83, 90, 95, 97, 106, 117, 120, 129, 130, 132, 133, 136, 137, 138; Bengal Sanitary Commission 60–62; Black Town 53, 61, 63, 64–65, 68–70; as city of beauties 57–58; as city of memories 74–77; as city of palaces 54–56; as city of pale faces 60, 63; Delhi (contrasted to) 72; *Englishman* 64–65; filth 60–65; Government House 54; Grey Town 54; history of 52–54; "little London" 55–56, 64; London (compared to) 55–56, 58, 63–66, 67, 68, 71; maidan 53, 54, 59, 60, 67, 72; as museum 68; as native metropolis 71–77; Old Park Street Cemetery 70–71; prostitution 68–70; *Report on the Royal Botanical Gardens and other gardens in Calcutta* 73–74; smell 64–67, 71; *The Statesman* 64–65; *Thacker's Guidebook to Calcutta* 64–65; waste disposal

(history of) 61–62, 63, 65; White Town 52–60, 61, 64, 65, 67, 68, 72
capitalism 3, 6, 37, 68, 151, 152
Carlyle, Thomas 91
caste 6, 11, 25, 57, 96, 134, 143, 151, 152; depiction by Chatterji, Bankim Chandra 97–98; depiction by Tagore, Rabindranath 103–104, 117, 128, 129, 130, 131, 132–137; depiction in *Kim* 120, 121, 122, 123; depiction in *The People of India* 121
Chakrabarty, Dipesh 4, 6, 43, 43n24, 43n25, 114n81, 115n83, 141; "asymmetric ignorance" 6; "waiting room of history" 43
Charnock, Job 52
Chatterjee, Partha 84, 43n24, 43n25, 108, 114n82, 142–143; spiritual 84, 108, 114n82, 142
Chatterji, Bankim Chandra 9, 10, 19; *Anandamath* 32–45; caste (depiction of) 97–98; Muslim (depiction of) 33, 34, 35, 41, 43, 44–45, 99, 138; *Vandemataram* 10, 32, 34, 39, 82, 83, 84, 85, 92–99, 100, 103, 104, 108
Chaudhuri, Supriya 128
chronotope *see* Bakhtin, Mikhail
Chunder, Bholanath 10; *Travels of a Hindoo* 72–73, 74, 75
city as *oeuvre see* Lefebvre, Henri
colonial globalization 3, 6, 7, 68, 150
constellation 5, 6, 9, 11, 19, 51, 84, 117
contradictory space (including spatial contradiction) *see* space
contrapuntal reading *see* Said, Edward
counter-pastoral 11, 104–105

Das, Sarat Chandra 126
Deb, Binay Krishna 10, 73
decolonization 1–2, 4, 5, 10, 11, 41, 77, 85, 102, 125, 143, 148–152; as event and process 149–150; failure of 151–152; and nationalism 10, 151; as space-making project 1–2, 85; as spatial desire 1–2, 5; as utopianism 4–5, 10, 11, 41, 105, 143, 148–149, 151–152
Dharma Sabha 129
Dickens, Charles 32, 68
disenchanted spaces *see* space
Durga *see* Hinduism

East India Company 20, 33, 34, 52–54, 56, 61, 71, 87, 88, 94
ethnography 2, 9, 11, 57, 66, 69, 76, 120, 121, 122, 126

famine (in Bengal and colonial India) 10, 33, 34, 36, 37, 38, 44, 84, 94
Fanon, Frantz 1
Farsi (Persian) 22, 26, 51, 57, 58, 59
filiation *see* Said, Edward
Forster, E. M. 28; *Passage to India* ("Marabar Hills") 28
Foucault, Michel 145n63
Frye, Northrop 18, 21

Gandhi, Mohandas 117, 134
Ganges, the (including The Hoogly) 8, 52, 56, 62, 65, 72, 76
gender 105, 117, 130–132, 151; depiction by Hossain, Rokeya 137–143; depiction by Tagore, Rabindranath 131–132, 137, 146n75; gendering of space 38–41, 83, 84, 87, 89–90, 92–100, 106–107, 109n5, 109n12; patriarchy 41, 117, 134, 139, 140, 141, 142, 143; and utopianism 41, 117, 131–132, 137, 138–143; "woman on top" 41; woman question 131, 142–143
geo-body 83, 103, 109n5
Ghalib, Mirza 56–58
globalectical reading 5–9
globalectics *see* Ngũgĩ wa Thiong'o
good and evil 9, 18–45
Goswami, Manu 14n24, 14n25
Grand Trunk Road 121–123
Great Cities of the World 59, 62–63
Great Game 89, 118, 125–126
Gyatso, Ugyen 126

Heber, Reginald 90–92
Heidegger, Martin 3; age of the world-picture 3
heteroglossia *see* Bakhtin, Mikhail
Hinduism 8, 39, 96, 97, 98, 107, 128, 129; caste 6, 11, 25, 57, 96, 97–98, 103–104, 117, 120, 121, 122, 123, 128, 129, 130, 131, 132–137, 134, 143, 151, 152; divine (conceptions of) 97, 107; Durga 38, 39, 41, 22, 83, 93, 94, 96, 97, 100; Jagaddhatri 39; Kali 39, 40, 41; liturgy 97–98; Mohishashur 41, 97; religious panegyrics 97, 98
homotopia *see* Mohan, Anupama
Hossain, Rokeya Sakhawat 11, 116, 117, 143; *Padmarag* 11, 116, 117, 137–143

idealization of space *see* space
idolization of space *see* space
imaginary resolution of real contradiction *see* Jameson, Fredric
imagination 2, 4, 8, 10, 11, 39, 45, 82, 83, 84, 86, 87, 90, 91, 96, 98, 99, 102, 103, 116, 123
imagined geographies *see* Said, Edward
internationalism 103
Islam 24, 83, 107, 134; *see also* Muslim

Jameson, Fredric 12n5, 15n33, 18, 21, 27, 32, 36, 37; affective impulse 27, 36; imaginary resolution of real contradiction 19, 32; romance 18, 21, 32, 37; semic evaporation 32

Kali *see* Hinduism
Kapila, Shuchi 23, 24
Kaul, Suvir 86, 87
Kaviraj, Sudipta 43, 44, 95, 98
Kipling, Rudyard 10, 11, 22, 52, 63, 65, 72, 75, 76, 116, 117, 129, 132, 136, 137; caste (depiction of) 120, 121, 122, 123; *City of Dreadful Night* 65–71; *Kim* 118–128; "Lispeth" 127–128; sexuality (depiction of) 127–128, 145n63; "The White Man's Burden" 91
Klein, Ira 94
knowable communities *see* Williams, Raymond
Kurtz 32

labour 2, 3, 4, 10, 11, 28, 38, 51, 53, 91, 95, 104, 123, 131, 139, 151
landscape 10, 27, 28, 33, 36, 37, 54, 59, 60, 61, 64, 65, 66, 67, 69, 71, 82, 84, 86, 87, 91–95, 97, 98, 101, 107, 110n28, 122, 123, 141, 142
Lazarus, Neil 4
Lefebvre, Henri 3, 10, 19; abstract space 3; city as oeuvre 10, 52; production of space 3
Lipner, Julius 98, 92–94
literary field 1, 2, 4, 5, 6, 7, 9, 32
Lukacs, Gyorgy [Georg] 43

Malamoud, Charles 97
Marx, Karl 2
masnavi 57, 79n32
Mearns, Andrew 63–64
metonym 7, 103, 108, 142; *see also* scale
Mohan, Anupama 44; homotopia 44
Montgomerie, Thomas 126
Muslim 10, 34, 43, 44, 45, 54, 83, 96, 99, 138, 143; *Anandamath* (depiction in) 33, 34, 35, 41, 43, 44–45, 99, 138; *Kim* (depiction in) 123, 126, 127; Hossain (depiction by) 117, 138, 141–142; Tagore (depiction by) 101–102, 103, 117, 128, 132–137, 142
mutiny novel 19, 19, 20, 21

Nandy, Ashish 129
narrative and space 2, 5–9, 12n9
National Anthem and National Song 10, 82, 83, 85
nationalism 4, 6, 11, 14n24, 14n25, 82, 83, 84, 90, 108, 117, 131, 142, 143, 151
nation-state 1, 82, 83, 85, 96, 151
Nehru, Jawaharlal 134; "tryst with destiny" speech 149–151
Ngũgĩ wa Thiong'o 6; globalectical reading 5–9
Not Yet *see* Bloch, Ernst
Not-Yet-Become *see* Bloch, Ernst
Not-Yet-Conscious *see* Bloch, Ernst

Pande, Mangal 21, 24
pastoral 10, 35, 91, 95, 104
patriotic lyric 9, 32, 39, 82, 84; anti-colonial version 92–108; colonial version 87–92; form and function 85–86; as global genre 10, 85–86
People of India, The 120–122, 126
Pinney, Christopher 121
place *see* space
place-making *see* space
postcolonial literature and theory 2, 4, 7, 117
postcolony 2, 4, 5, 9, 10, 11, 19, 35, 37, 45, 82, 84, 85, 86, 102, 103, 116, 117, 138, 141, 143, 151
Prasad, Rajendra 82
production of space *see* Lefebvre, Henri
progressive history 39, 41, 43–44, 103

Ramazani, Jahan 102
Rawat, Nain Singh 126
Ray, Dwijendralal 99–100, 113n64
realism 27, 37, 38, 132, 133; affective impulse 27, 36; Tagore 146n82
re-enchantment of space *see* space
romance 9, 18–21, 23, 32, 33, 36–38, 43–45, 132

Said, Edward 1, 2, 47n19, 83, 118, 145n42; affiliation 84, 100, 102, 103; contrapuntal reading 5–6; filiation 84,

92–99, 102, 103; imagined geographies 1, 2, 8, 9, 12n4, 19, 27, 34, 116, 132, 143; narrative and space 2, 12n9
Sarkar, Sumit 72, 109n10
Savarkar, Vinayak Damodar 7–9
scale 2–4, 10, 11, 52, 77, 103, 130, 131, 142; *see also* metonym
Scott, Walter 33
semic evaporation *see* Jameson, Fredric
Sepoy Mutiny *see* 1857 Uprising
sexual desire 24, 127–128, 145n63
shahrashub 57–58
Sharma, Sunil 57
simultaneity of the non-simultaneous *see* Bloch, Ernst
Siraj-ud-daulah 33, 52
Smith, David 61, 72
Smith, Neil 3
space: anthropomorphic representation of 38–41, 84, 87–100; border (including boundary) 3, 151; contradictory space (including spatial contradiction) 27, 28, 35–37, 52, 56, 64, 65, 116, 118–128, 136; disenchanted spaces 4, 38, 86, 98, 118; gendering of space 38–41, 83, 84, 87, 89–90, 92–100, 106–107, 109n5, 109n12; geo-body 83, 103, 109n5; idealization of 29, 30, 84, 90, 93–96, 134; idolization (deification) of 38–41, 84, 87, 89–90, 92–100; landscape 10, 27, 28, 33, 36, 37, 54, 59, 60, 61, 64, 65, 66, 67, 69, 71, 82, 84, 86, 87, 91–95, 97, 98, 101, 107, 110n28, 122, 123, 141, 142; place 27, 28, 37, 71, 85, 86, 98, 116–143; place-making 2–4, 150; postcolony 2, 4, 5, 9, 10, 11, 19, 35, 37, 45, 82, 84, 85, 86, 102, 103, 116, 117, 138, 141, 143, 151; re-enchantment of 4, 10, 38, 86, 98; space-making 2–4, 150; spatial desire 1–12, 18, 19, 23, 28, 29, 30, 35, 37, 38, 43, 44, 45, 52, 77, 84, 86, 98, 100, 102, 103, 117, 125, 128, 132, 143, 148–152; spatialized alienation 77; spatialized heteroglossia 11, 117, 128–137; spatialized unevenness 51; world-making 3
space-making *see* space
spatial contradiction *see* space
spatial desire 1–12, 18, 19, 23, 28, 29, 30, 35, 37, 38, 43, 44, 45, 52, 77, 84, 86, 98, 100, 102, 103, 117, 125, 128, 132, 143, 148–152
spatialized alienation *see* space
spatialized heteroglossia *see* space
spatialized unevenness *see* space
spiritual 84, 87, 97, 98, 100, 101, 107, 108, 114n82, 142
Spivak, Gayatri 30
Strachey, John 8–9, 60–61, 62, 63, 136
structure of feeling *see* Williams, Raymond

Tagore, Rabindranath 7, 8, 9, 10, 11, 17n48, 52, 99, 117, 128, 136, 137, 146n82; caste (depiction of) 103–104, 117, 128, 129, 130, 131, 132–137; class (depiction of) 102–104, 132–137; gender (depiction of) 131–132, 137, 146n75; *Gora* 4, 116–117, 128–137, 140, 142; *Home and the World* 109n10, 146n75, 147n83; *Jana Gana Mana* 10, 82, 84, 85, 100–102, 105–108, 154–156; *Jana Gana Mana* (controversy) 106–107; *My Boyhood Days* 74–77; Muslim (depiction of) 132–137; "Pilgrimage to India" 102–104; relationship with *Vandemataram* 83, 95, 99, 109n10, 113n63; spatialized heteroglossia 11, 117, 128–137
Taylor, Philp Meadows 9, 19; Seeta 9, 19–32
Telscher, Kate 55
Thomson, James 10, 85, 87, 89, 90, 94, 100; "Rule! Britannia" 10, 85, 87, 89–90, 94, 100
Tibet 118, 125, 126
translatio imperii 151

utopia (including utopianism) 1–2, 4, 9, 11, 19, 43, 103, 105, 106, 108, 116, 117, 129, 130, 131, 135–137, 142, 148–152; as aesthetic education 5, 86; as collective dreaming 4; colonial utopianism 1–2, 20–32; and decolonization 4–5, 11, 41, 148, 152; as determinate negation 10, 95; as education of desire 4–5; Ernst Bloch 5, 7, 19, 114n82, 148; and gender 41, 117, 131–132, 137, 138–143; as impossible space 36–39; non-western utopia 6, 15n40; postcolony 2, 4, 5, 9, 10, 11, 19, 35, 37, 45, 82, 84, 85, 86, 102, 103, 116, 117, 138, 141, 143, 151; in post-independence India 11, 149–151, 152; and romance 10, 18–19, 20–45; spatial desire 1–12, 18, 19, 23, 28, 29, 30, 35, 37, 38, 43, 44, 45, 52, 77, 84, 86, 98, 100, 102, 103, 117, 125, 128, 132, 143, 148–152; as structure of feeling 1–2, 37–39; as subjunctive 132, 133
utopianism *see* utopia

Williams, Raymond 2, 117; knowable communities 117; landscape 95, 110n 28, 122; narrative and space 2, 12n9; pastoral 95; structure of feeling 2, 12n5, 37, 151
Winichakul, Thongchai 3, 109n5
woman question 131, 142–143
world-making *see* space

Younghusband, Francis 126